THE PRIVATE EQUITY AI OPERATING PARTNER

THE PRIVATE EQUITY AI OPERATING PARTNER

USING ARTIFICIAL INTELLIGENCE FOR VALUE CREATION

Bruce Sinclair

Founder of aioperatingpartners.ai

EVE ALLAN

Library of Congress Control Number is available upon request.

ISBN 979-8-218-83986-4

Printed in the United States of America
10 9 8 7 6 5 4 3 2 1

Version 1.

With love to Sarah, Paris and Chase.

CONTENTS

ACKNOWLEDGMENTS

> The answer is out there, Neo, and it's looking for you, and it will find you if you want it to.
>
> —*Trinity*

Every book has its origin story, and for this book it's been percolating my entire career. Although I first started writing it 10 years ago under the title *AI Inc* (as a follow-on to my first book, *Iot Inc.*), I've been preparing to write about AI since the beginning. As a mathematician and programmer, I understood it. As the CEO of an AI company, I sold it. As an operating partner and consultant, I deployed it. But it was only last year when I came to the realization that since I've been using AI to create value the whole time, I should share my thoughts about it from that perspective. AI may have been called other names as part of the internet of things (IoT) and digital transformation, but it's always been the creator of value—the thing that transformed data into useful information. As a capstone I wanted to pull together everything I knew about the subject, stitch it together for business and investing and release it into the world. And one year later, now that it's finished, I feel complete. Getting me to this place has been the result of many people whom I'd like to acknowledge.

I'd first like to thank Rajeev Amara who had the vision to recruit me from consulting to join Arcline's value creation team as a digital operating partner. It was the experience I had with the other partners I worked with and our portfolio companies that set my career in a different direction.

I'm grateful to Keng Ooi Tan and Matthew Swanson who as recent clients (and subsequent friends) were "all in" on incorporating Analytical AI and Generative AI into their respective businesses.

I'm also thankful to my peer reviewers for their early interest in the book and for their time in reviewing their parts—making the whole better. Specifically I'd like to thank Arun Karthik Ganesan, John Hajdukiewicz, Chase Sinclair, Daryl Moon, Shiva Sawant, Kevin Jones, Varun Singal, Doug Razzano, Barry Low, Michael Mihuc, Samuele Da Ros, Matthew Swanson, Garret Twining and Dan Seitam.

I'm most appreciative of my loved ones—family and friends—who each supported me in their own way during the writing of this book.

INTRODUCTION

WHAT'S THE DEAL WITH AI?

AI has been around for a while. My first professional exposure to it was as the CEO of an AI company years ago when I lived in Montreal. Our AI simulated the movement of thousands of digital extras for movies and non-player characters (NPCs) for video games.

There are many perspectives from which to look at AI. I first looked at it to create visual effects; now I look at it to create enterprise value.

WHAT THIS BOOK IS ABOUT	WHAT THIS BOOK IS NOT ABOUT
✔ AI Strategy	✘ Operational Theory
✔ AI Risk Mitigation	✘ Change Management
✔ AI Value Creation	✘ Justification
✔ Getting Started	✘ AI History/Philosophy
✔ How To	✘ Academic
✔ First Principles	✘ Case Study Bingo
✔ Portfolio Improvement	✘ Internal PE Operations

This book, however, doesn't spend your time on operational improvement theory or change management or on trying to convince you of the importance of value creation in private equity (PE) today by quoting surveys or studies. Nor does it evangelize AI or explore its history or the philosophical questions it invokes.

Far from academic or preachy, this is a how-to book, with the "hows" gained from years of applied AI experience. And to make it practical, it's written with the fewest pages and the most case studies and examples—almost all from our past clients. These are case studies to reuse, leverage or riff off for your portfolio.

Pattern matching case studies of what's been done before to your business needs is a common way to get started, but this book doesn't stop there. It also equips you to innovate by distilling the complex subjects of AI and value creation into the first principles needed to bridge these two worlds so you can make them work together in your own original way.

If it's not clear by now, this book does not cover how to use AI internally to improve firm operations—instead it guides you on how to use Generative AI (GenAI) and Analytical AI to improve portfolio companies.

QUESTIONS THIS BOOK ANSWERS

This book answers your most important "How do I" questions:

1. *How do I develop my AI strategy?*
2. *How do I identify AI risk?*
3. *How do I create value with AI?*
4. *How do I quantify AI value?*
5. *How do I execute my AI strategy?*

Notice from this line of questioning that strategy is just the beginning. This book provides a playbook for realizing company-level strategy by showing you how to create value with AI and how to best interface with technical partners to implement it.

From here we'll have everything needed to answer the bonus question:

6. *How do I get started?*

This book takes you top to bottom, from AI strategy down to its bits and bytes, to answer these six questions by using my original playbook and tools.

WHO IS THIS BOOK FOR?

This book is for any general partner (GP) or senior executive who wants to harness the power of AI. It's also for limited partners (LPs) who are learning the AI value landscape and for private equity service providers in banking and consulting who need to understand AI to effectively serve their clients.

WHY YOU NEED TO READ THIS BOOK

Every Company Needs an AI Strategy

AI is here and getting more useful by the day. It's a broad horizontal layer able to improve every part of every company. Every deal should be considered from an AI point of view.

The GenAI eruption has been a wake-up call letting us know "there's a there there." But most companies aren't using AI beyond prompts, or if they are, they're using it for an engineering project, not for an investment. And that's a recipe for disaster. And that's why a vast majority of AI projects fail: because there isn't a value creation strategy behind them.

You Need to Know How AI Works in Order to Put It to Work

Because of AI's strategic nature, it's not something you can completely delegate to others. Therefore, to give you the knowledge you need to incorporate AI into your planning, we will be discussing AI from value creation, risk, financial and technical perspectives.

This book will provide you with the AI literacy you need to thoughtfully discuss AI use with all your stakeholders. Not enough to code it yourself, but enough to hire and manage the tech partners who will do so on your behalf. And enough to work with boards and leadership to help them think about AI strategically and out of the box.

The Book Will Give Your Firm a Competitive Edge

Applying this hands-on AI playbook with its original model building blocks, value creation frameworks, risk/reward assessments, value levers, development workflows and execution plans will equip you and your firm with a competitive edge over firms that ignore AI or use it ineffectively.

The Takeaway

After reading this book you'll be able to look at companies through the PE AI lens, pre- and post-acquisition, and see value creation opportunities that may be invisible to your competitors. Equipped with this knowledge, your firm will have the conviction to bid higher to win more auctions, increase the value of your portfolio in novel ways, and side-step AI risk.

AI IS NOT JUST FOR TECH SECTORS ANYMORE

AI has the potential to create value across every sector, from digital native firms to tech-enabled service providers and with digital twin technology (software to simulate physical value creation), to traditional, non-technical real-world businesses.

TERMINOLOGY USED

AI: Generally, AI is used as an abbreviation for an AI application or AI agent that is utilized stand-alone or embedded into a company's products, services, places or operations. It's also an umbrella term for GenAI and all the flavors of Analytical AI, including symbolic AI, machine learning and neural networks.

PE AI: Private equity AI is artificial intelligence used to increase the performance of private equity firms and the enterprise value of their portfolio companies.

AI Operating Partner: The AI operating partner is generalized in this book to mean a full-time or fractional individual or team that helps PE firms and their companies deploy PE AI.

Product: A product in this book is shorthand for a product, a service or a place.

BOOK OVERVIEW

This book is divided into three parts.

Part One: Strategic AI Value Creation

This part of the book bridges value creation and risk assessment with strategy. It starts with first principles by applying what AI can do to answer the biggest of questions: *How do I create value with AI?*, *How do I identify AI risk?*, and ultimately, *How do I develop my AI strategy?* It includes examples of each AI ability and a detailed case study from an early client using the AI Value Creation Framework.

Part Two: AI Value Drivers

This part of the book bridges value creation with finance. It goes through the three value drivers and each of the 10 AI value levers to demonstrate another way to imagine AI value creation. This approach starts by classifying value as growth, cost reduction or multiple expansion, enabling us to answer the question, *How do I quantify AI value?* A reusable client case study is included for each value lever.

Part Three: Executing AI Strategy

This is the techie part of the book. It answers the question, *How do I execute my AI strategy?* by introducing the AI Tech Stack to explain how AI works and the AI Requirements Doc for communicating exactly what needs to be developed by the tech team, bridging the age-old chasm between strategy and execution. It ends with a detailed example of AI math and algorithms.

Conclusion: Getting Started

The book's conclusion brings it all together by answering the question, *How do I get started?*

• • •

Artificial Intelligence is a generational tech that can improve every aspect of business—so let's put it to work to create value! We start with the fundamentals by learning what AI is capable of.

PART ONE

STRATEGIC AI VALUE CREATION

This book is a formalized version of the playbook I've been using over the last 10 years. Its North Star is called the AI Value Chain, which we'll be using throughout the book to ensure value creation is the guiding light for everything we do.

AI VALUE CHAIN

STRATEGY
VALUE
USEFUL INFO
AI MODEL
DATA
SOFTWARE
HARDWARE

We use the value chain to develop our AI strategy and establish a framework and workflows to help us define, quantify and cost AI value creation.

Going down the chain establishes requirements. Starting at the top, our strategy defines the requirements for the value we create, which defines the requirements for the useful information we must produce, and so on, all the way to the hardware, which is chosen to support the requirements of the software and, by definition, everything else above it.

It's important to note that figuring out the tech is at the end. A rookie mistake is to start with the tech and expand from there.

Going up the chain establishes the dependencies. Starting at the bottom, our hardware is dependent on the needs of our software, which is dependent on the needs of our data, and so on all the way to AI strategy, which is dependent on company strategy and the investment thesis or value creation plan.

The AI Value Chain shows the correct order of things: the order of the steps we take, and the relationship between the steps along the way to navigate the path between strategy, value creation and AI.

• • •

Although our North Star is strategy, we start with the way-points that will lead us there. In Chapter 1 we'll see that all AIs can be broken down into combinations of AI abilities. Chapter 2 will introduce the AI Value Creation Framework, and Chapter 3 will cover AI risk mitigation. Chapter 4 will build upon the first three chapters to lay a course to AI strategy. Chapter 5 is about AI applications and agents—what we use to create value and realize our strategy. All right? Let's go!

CHAPTER 1

FIRST PRINCIPLES OF APPLIED AI

AI VALUE CHAIN

STRATEGY
VALUE
USEFUL INFO
AI MODEL
DATA
SOFTWARE
HARDWARE

The AI model is where the action is. What we can do with AI comes from what we can do with the AI model. It's the brains of the operation and sits squarely in the center of the AI Value Chain. In this chapter we discuss what AI models and, by extension, AIs are able to do.

The AI world of the possible is unbounded and can be highly complex.

While pattern matching to a list of AI use cases is an easy way to dip your toes into AI, it's not what I cover here. AI use-case bingo can be found everywhere, including from your favorite GenAI chatbot. Although a good first step, simply showing a myriad of examples to choose from eliminates originality and only scratches the surface of what's possible because use cases sidestep why things work the way they do.

In this chapter we're going to address why AI works the way it does. The complexities of AI are simplified into abilities that are indivisible by other abilities. To conceptually understand what we can do with AI, we employ first principles. We begin by introducing the 12 Atomic AI Abilities:

The 12 Atomic AI Abilities

1. CLASSIFY
2. GROUP
3. IDENTIFY
4. DETECT ANOMALIES
5. ASSOCIATE
6. RECOMMEND
7. INTERPRET
8. LEARN
9. SIMULATE
10. OPTIMIZE
11. PREDICT
12. CREATE

In combination, these AI abilities practically describe what we can do with AI.

All AI value comes from transforming data into useful information.

The Atomic AI Abilities are the transformers, so understanding them is important for two reasons:

1. Combining these abilities in different ways allows you to use AI to innovate with existing products (services, places, operations) or invent new ones or to make products more efficient.
2. Specifying which abilities to use along with the input (data) and the output (useful information) allows you to clearly specify your AI requirements to your tech partners. The better you can speak the language of the data science team implementing your requirements, the more efficient their work will be—resulting in higher-quality models in less time.

ATOMIC AI ABILITIES

This list of the 12 Atomic AI Abilities is an important new vocabulary for AI value creation practitioners to learn.

CLASSIFY: Sort into predefined categories.
Classify things by trying all the combinations of how each thing is like each category.

- Example: Classify medical images to assist radiologists in diagnosing diseases.

GROUP: Sort into clusters based on similarity.
Group things by trying all the combinations of how each thing is like each other.

- Example: Group shopper behaviors to improve store layout and signage.

IDENTIFY: Recognize objects or patterns.

Identify things by matching them to other things that were identified in the past.

- Example: Computer vision inspection to find defects on assembly lines.

DETECT ANOMALIES: Identify outliers.

Find things that aren't like what's been there in the past.

- Example: Monitoring to detect structural irregularities in bridges and tunnels.

ASSOCIATE: Link related data together.

Associate things to each other based on proven connections in the past.

- Example: Sentiment analysis linking buying decision to guide product development.

RECOMMEND: Suggest actions or items.

Recommend things based on what was successfully chosen in the past.

- Example: Contract analysis tool that suggests negotiation strategies.

INTERPRET: Find meaning or insights.

Interpret things by making associations between similar things that happened in the past.

- Example: Analysis tool that extracts insights from IT system logs to prevent downtime.

LEARN: Acquire/refine knowledge or skills.

Learn things by pattern matching to similar things learned in the past.

- Example: Risk assessment model using proprietary data to underwrite insurance.

SIMULATE: Create and use a stochastic digital representation.

Simulate things by matching the patterns of how things behaved in the past.

- Example: Airport simulation to plan traffic and pedestrian flow.

OPTIMIZE: Find the best solutions for a goal.

Optimize things by trying a bunch of solutions to see which one's best.

- Example: Optimize inventory distribution to reduce shipping costs and delays.

PREDICT: Forecast outcomes.

Predict things based on what has happened in similar situations in the past.

- Example: Predictive maintenance to reduce down-time.

CREATE: Create new content.

Create new things based on what has been created in the past.

- Example: Improve machine documentation by making it prompt-based.

On a technical note, each atomic ability corresponds to a category of commercial/mathematical AI models that have the same abilities but solve the same problem type in different ways. (See Appendix 1 at the end of this chapter.) The specific mathematical model chosen by the data scientist to implement an AI ability specified here depends on the available data and computing resources.

AI ABILITIES "PERIODIC TABLE"

To help you better understand how the Atomic AI Abilities are related to each other, they are presented like the elements in a periodic table with the columns representing a "family" of AI abilities and the rows representing increasing levels of sophistication (see Table 1).

Table 1 **Periodic Table of AI Abilities**

FAMILY	PERCEIVE Senses and structures data	REASON Analyzes patterns and extracts meaning	ACT Decides, forecasts and generates outputs
ABILITIES	CLASSIFY (Sort Data)	ASSOCIATE (Link Data)	SIMULATE (Model)
	GROUP (Cluster Data)	RECOMMEND (Advise)	OPTIMIZE (Refine)
	IDENTIFY (Recognize)	INTERPRET (Explain)	PREDICT (Forecast)
	DETECT ANOMALIES (Outliers)	LEARN (Adapt)	CREATE (Generate)

The AI Abilities "Periodic Table" provides a structured way to compose AI solutions to value creation problems by stringing together one or more AI abilities. This table and the previous AI abilities list are two handy resources you'll refer to when imagining your AIs.

Stringing AI abilities together as an AI application is analogous to a tool: There is one input, one or more AI abilities executed and one output:

Input → AI abilities → Output

Stringing them together as an AI agent is analogous to adding a teammate. But instead of a linear progression, it's a loop. A loop that continuously executes one or more PERCEIVE AI abilities, then one or more REASON abilities and then one or more ACT abilities and then loops back to PERCEIVE:

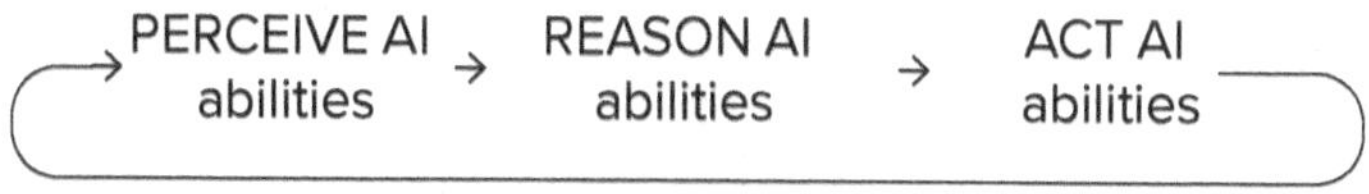

More about how we package AI abilities as AI applications and AI agents will be discussed in Chapter 5.

USING AI ABILITIES

This is where imagination comes into play. Think of the Atomic AI Abilities as 12 differently shaped Lego blocks. Like with Lego, the innovation and invention come from how the blocks are assembled to realize your vision. And importantly, if you can describe your vision or solution to create value as a combination of the 12 abilities, they can be realized with AI.

Applying them means we need to understand the science, technology, engineering and mathematics (STEM) behind how we're going to create value. Often this means teaming with subject matter experts to break down and formulate how value is being created in the form of AI abilities.

Showing 'em in Action

To find our AI value creation solution, we string together one or more AI abilities to deliver the targeted useful information.

Example 1: Track inventory levels via sensors (IDENTIFY), analyze trends in sales and supply chain data (INTERPRET), forecast future demand fluctuations (PREDICT), and then determine the best restocking schedules and reorder points (OPTIMIZE) to reduce stockouts and overstocking.

Example 2: CLASSIFY product returns by defect, INTERPRET insights into causes, and then PREDICT future return rates for just-in-time manufacturing

Example 3: CLASSIFY → INTERPRET → PREDICT for customer churn forecasting.

Example 4: IDENTIFY → ASSOCIATE → OPTIMIZE for supply chain logistics improvements.

Note that in these four examples, multiple AIs are needed for each solution. Since each AI ability is realized by a separate AI, most solutions require multiple AI models stitched together with software. More on this in Chapter 5.

MORE ATOMIC THOUGHTS

As you will learn in Part Two, if you squint your eyes, you'll see that AI value creation comes down to innovation, invention and efficiency.

AI innovation lies in how you break down value creation problems into these AI abilities. AI invention occurs when you assemble these abilities in new ways to create value. And AI efficiency happens when you teach AI to do better what's currently being done.

The Atomic AI Abilities are the building blocks used to imagine novel AI solutions by pinpointing the kind of "intelligence" AI models need to transform data into useful information.

• • •

Pretty heady stuff. Now in Chapter 2 let's see how AI abilities are used in the context of the AI Value Creation Framework.

APPENDIX 1: TYPES OF AI MODELS

This super-techie table identifies the commercial or mathematical names of the AI models sold and used today as related to the Atomic AI Abilities. While GenAI models have been on everyone's mind and are even sometimes synonymous with AI, this table demonstrates the rich ecosystem of other AI model types that are often a better value creation solution.

GenAI models can be used to do pretty much all 12 of the Atomic AI Abilities under the right circumstances, but they are often way overkill. If GenAI is suggested for your value creation needs, consider the Analytical AI model types too. Because they are specialized, they can be cheaper, more efficient and easier to implement than general GenAI models that are expensive to train, require massive amounts of data and resources (compute and memory), encroach on data sovereignty and have been known to hallucinate (to get things wrong). Don't get me wrong, I love my GenAI, but even though GenAI may be able to solve your problem, it may not be the right or only solution that can.

Atomic Ability	Model Types	Family	Training
CLASSIFY	Logistic Regression, Decision Tree, Random Forest, SVM	Tree-Based, Linear Models	Supervised Learning
GROUP	K-Means, Gaussian Mixture Models (GMM), Principal Component Analysis	Clustering Models	Unsupervised Learning
IDENTIFY	Convolutional Neural Network (CNN), Object Detection Models	Neural Networks	Supervised/Deep Learning
DETECT ANOMALIES	Isolation Forest, Autoencoder, One-Class SVM	Anomaly Detection Models	Unsupervised/Semi-Supervised/ Deep Learning
ASSOCIATE	Graph Neural Networks (GNNs), Bayesian	Neural Networks	Unsupervised/Probabilistic Learning
RECOMMEND	Collaborative Filtering, Matrix Factorization, Content-Based Filtering	Recommendation Models	Supervised/Unsupervised Hybrid
INTERPRET	GenAI: Multimodal Reasoning, Chain-of-Thought (CoT), Neuro-Symbolic	Reasoning Models	Supervised/ Reinforcement Learning

(continued on next page)

Atomic Ability	Model Types	Family	Training
LEARN	Autoencoder, BERT, Contrastive Learning	Self-Supervised Models	Unsupervised
SIMULATE	Agent-Based Modeling, GANs, Markov Models, Q-learning	Simulation Models	Unsupervised/ Reinforcement Learning
OPTIMIZE	Linear Programming, Genetic Algorithms, Bayesian	Optimization Models	Optimization/Mathematical Programming
PREDICT	Bayesian, Prophet, XGBoost	Forecasting Models	Supervised Learning, Time Series Forecasting
CREATE	GenAI: Generative Adversarial Network, Diffusion	Generative Models	Supervised/Generative/Deep Learning

CHAPTER 2

THE AI VALUE CREATION FRAMEWORK

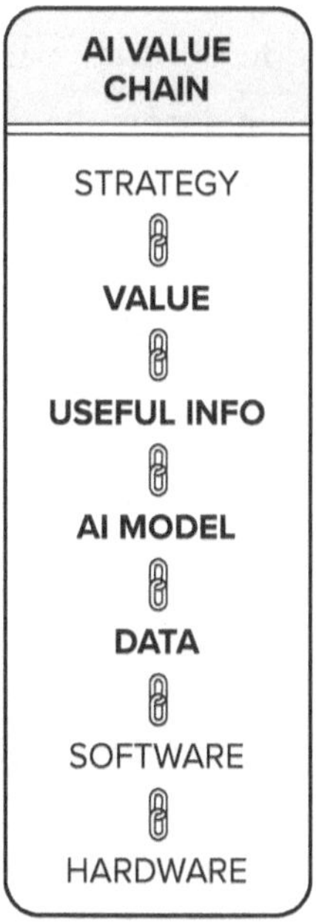

The hands-on AI journey for most companies goes bottom-up by starting in engineering with a POC (proof of concept). Techies love POCs almost as much as they love their tech. And with the best of intentions, they will learn Analytical AI or GenAI and then look for a problem to solve with it.

That's the tail wagging the dog, and it's exactly why a vast majority of AI initiatives fail: because they are tech-driven rather than value-driven.

To avoid the POC trap where time and money disappear and to align value creation with strategy, we work top-down within the AI Value Creation Framework.

This framework is about how to create value, not about what value to create. "What" is the work of strategy, which is covered in Chapter 4.

The value creation framework takes a step-by-step approach to creating value with AI. We start by defining the value proposition of the AI that's being developed. That value must be aligned with our strategy and defines the type of useful information we need, which in turn defines the requirements for the AI model and the data needed. Notice that we only start thinking about the tech at the end.

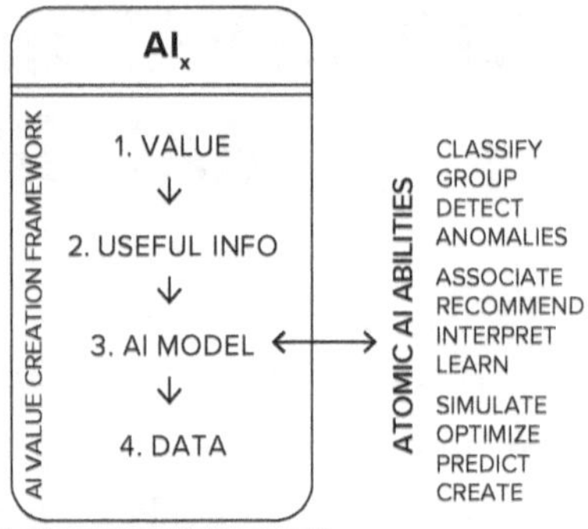

AI_x refers to a generic AI app or agent or more literally, the xth AI app or agent.

As part of step 3, we consider the Atomic AI Abilities covered in Chapter 1. These AI abilities define how the model will transform the data into useful information.

USING THE FRAMEWORK STEP BY STEP

When using the AI Value Creation Framework, we **align** each step with the step before and **articulate** the essence of the current step.

Let's go through the framework, step by step.

Step 1: Define the Value Proposition

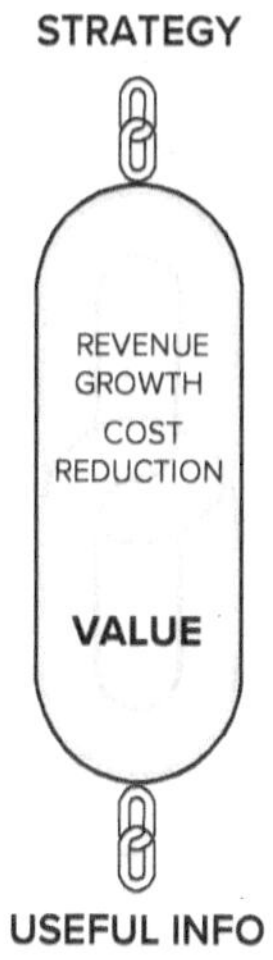

Align the value proposition with the end customer's needs by asking, *What outcome do we want to achieve, and how does it support the strategy?* The value produced by AI is qualified by the value proposition and quantified by classifying and then financially modeling it as discussed in Part Two.

Ultimately, what we're trying to do is increase the enterprise value of the company. The value we produce supports the strategy and defines the useful information needed.

The value proposition defined is for the end user of the AI. This means that the value prop can be for the company's end customer if the AI is to grow the company's revenue (through market share expansion, customer expansion, product expansion, market expansion, customer retention or pricing optimization), or the value prop can be for the company itself if the AI is to reduce costs (through margin improvement or operational efficiency). And the value prop is again for the company, sometimes by way of the end customer, when expanding the company's multiple (through add-on acquisitions or an AI rerate).

Articulate the value proposition and the business value it produces.

Step 2: Identify Useful Information

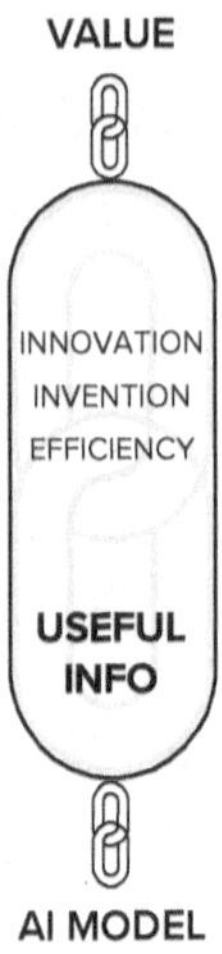

Align the information produced with the value proposition by asking, *What information would help realize this value?*

The useful information is what's valuable. It enables the value proposition and defines the output of the AI model. It's the information that innovates existing products (services, places) or invents new ones. Or it's the information that makes operations more efficient.

The useful information can be in the form of business intelligence to enable better decision making, but more often it is inputted for use by other AI or software.

Articulate the useful information to be produced and how it will be used.

Step 3: Define the Requirements of the AI Model

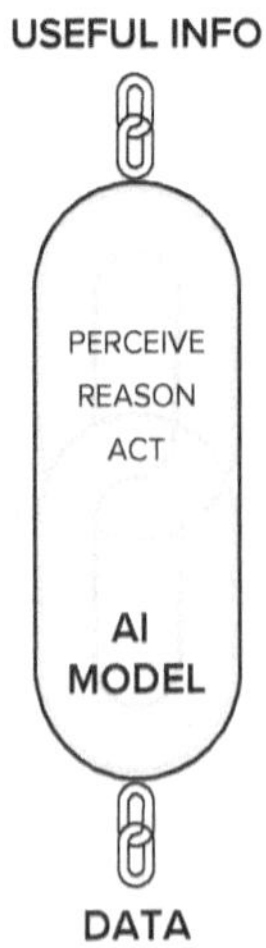

We spend an extraordinary amount of time in this book learning about the AI model because it is the center of all value creation. First in Chapter 1 to understand its capabilities (the ways it transforms), here to learn how it's defined and then in Chapter 5 to understand how it's packaged and in Chapter 10 to understand technically how it works. Why? Because:

All AI value comes from transforming data into useful information.

Align the AI model's abilities with the information to be produced by asking, *What does the model need to do to output the useful information?* Or more specifically:

Does the AI need to PERCEIVE?

If you have unstructured or raw data that needs processing, start with CLASSIFY, GROUP, IDENTIFY or DETECT ANOMALIES.

Or does the AI need to reason?

If you need to extract insights or recommendations or build a knowledge base, consider ASSOCIATE, RECOMMEND, INTERPRET or LEARN.

Or does the AI need to act?

If you need to model different scenarios, optimize for a particular outcome, predict future events or generate new content, use SIMULATE, OPTIMIZE, PREDICT or CREATE.

Articulate the AI ability that should be used from each selected family.

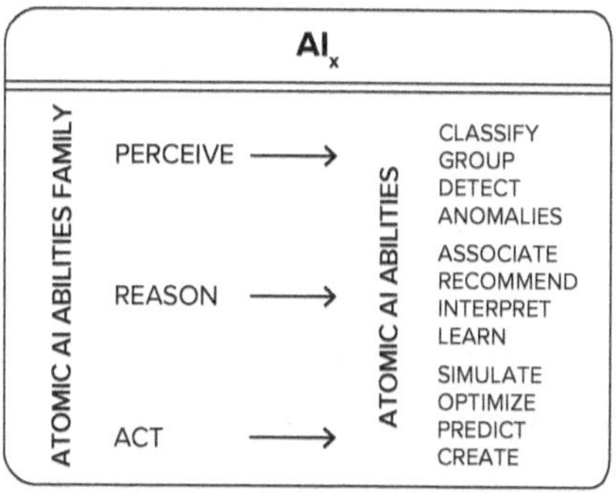

Here we choose the atomic AI ability from the ability family that best describes what needs to be done. From PERCEIVE we choose from CLASSIFY, GROUP, IDENTIFY or DETECT ANOMALIES. From REASON we choose from ASSOCIATE, RECOMMEND, INTERPRET or LEARN. From ACT we choose from SIMULATE, OPTIMIZE, PREDICT or CREATE.

Step 4: Find Data

Align the data that needs to be collected for the AI by asking, *What data do we need to feed into the AI in order to output the useful information? How do we know what data we need?* The AI model defines the requirements of the data it needs to produce valuable information. And more times than not, this data is not just sitting somewhere waiting to be tapped. It must be put in place.

Like the proprietary data captured by Amazon to make recommendations, the data needs to be proprietary to produce a competitive advantage; otherwise any other company could do the same thing.

This proprietary data is first used to train the AI model. Once trained, future data of the same type is inputted into the model and useful information is outputted to create value. This data is sourced from internal systems, external services and the internet of things.

Articulate how the data will be collected.

Once you get the hang of it, this is just a new way of thinking. Let's practice with a fully fleshed-out case study. Enter the road roller.

VALUE CREATION FRAMEWORK CASE STUDY

To ground this framework, we'll walk through a case study based on what I developed for Autodesk, an early client. The case study I'll share here and expand upon later in the book was developed for one of Autodesk's OEM customers. It's kind of fun; it involves a road roller, aka steam roller, squishing things.

Road stiffness and durability are related to asphalt density. Every geography, due to environmental factors, has a unique and ideal asphalt density to make the perfect road.

When paving a road, the job of the road roller is to compress the hot mixed asphalt, or HMA, laid by the paver, to the ideal asphalt density.

Road rollers use the weight and vibration of their drum to press air out from the hot mixed asphalt to increase its density.

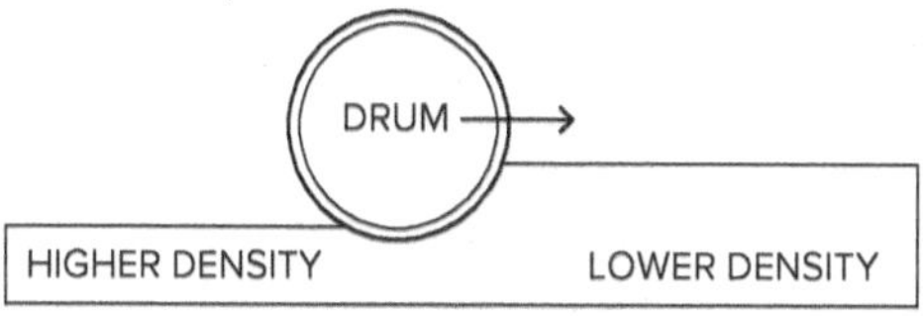

Let's walk through each step of the AI Value Creation Framework using the road roller to demonstrate how the OEM achieved its strategy.

Step 1: Define the Value Proposition

Align

The road roller OEM's strategy was to expand market share by developing a product that reduced the time and cost of paving roads.

Articulate

The value proposition for the OEM's end customer was, "Compress asphalt to the perfect density in the shortest amount of time." This enabled the end customer to reduce fixed costs (margin improvement) and to offer a more competitive paving service (market share expansion).

Step 2: Identify Useful Information

Align

Since the road roller's value proposition was about reducing paving time, the useful information has to be related to the road roller's speed.

Articulate

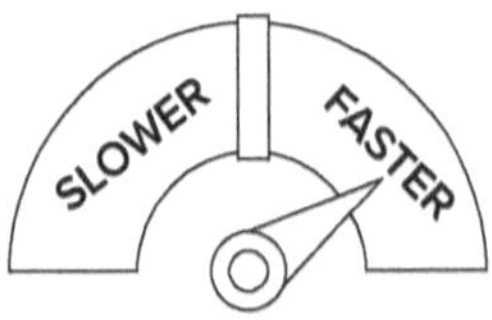

Given a specified asphalt density, the useful information was the road roller's max speed. This information was communicated with a speed gauge. A slower/faster meter was installed to instruct the operator to speed up or slow down. Although autonomous control of roller speed was possible, it was rejected at the time.

Step 3: Define the Requirements of the AI Model

Align

The model we developed for the road roller wasn't about the road roller, but about how the asphalt it rolls over is compressed. To model this we learned the physics of asphalt compression, which resulted in the following high-level function:

Asphalt density = *f*(HMA temp, drum weight, drum frequency, roller speed)

The output of our AI model needed to be roller speed, so the function had to be rearranged.

Since every geography has a known ideal asphalt density, we can make asphalt density a constant and reorder the function to solve for the road roller speed that achieves the desired compression by using the speed function:

Speed = *f*(desired asphalt density, drum weight, drum frequency, HMA temp)

Now let's look inside these functions from an AI perspective.

The first function, asphalt density, is a simulation of how the hot mixed asphalt is squished (SIMULATE).

The second function, for speed, is an underconstrained maximization problem (OPTIMIZE). The AI tries different values until it converges on the best answer.

Articulate

Along with identifying the useful information to be outputted (step 2) and finding the data (step 4), the requirements for the AI model are SIMULATE for asphalt density and OPTIMIZE for speed:

SIMULATE → OPTIMIZE

Step 4: Find Data

Align

The data we need to feed into the road roller AI model(s) includes:

- Desired asphalt density—the ideal density based on the environmental conditions where the road is being paved.
- Drum weight—the force applied by the roller's drum.
- Drum frequency—the frequency of the roller drum's vibration.
- HMA temp—the temperature of the hot mixed asphalt.

Articulate

The data will be collected in the following way:

- Desired asphalt density—asphalt engineering tables or, better yet, the company collects this data locally, which will inevitably be more accurate than public generic values.
- Drum weight—road roller specifications.
- Drum frequency—road roller specifications.
- HMA temp—IoT sensors on the road roller.

CASE STUDY SUMMARY

The OEM's AI strategy was to innovate its road roller product to grow revenue, categorized by the market share expansion value driver. Using the AI Value Creation Framework results in:

ROAD ROLLER AI

VALUE
End customer: Increase margins by compressing asphalt to the perfect density in the shortest amount of time.

↓

USEFUL INFO
The fastest speed the road roller can move while maintaining the desired asphalt density.

↓

AI MODEL
SIMULATE → OPTIMIZE
SIMULATE the hot mixed asphalt; then OPTIMIZE the equation solving for speed.

↓

DATA
Desired asphalt density (table), drum weight (specs), drum frequency (specs), HMA temp (IoT).

• • •

Need more examples? In Part Two of this book, we'll go through 10 more client case studies to further demonstrate how to use the value creation framework.

CHAPTER 3

AI RISK ASSESSMENT

If you listen to some AI pundits talk, it's easy to be low-key concerned about all PE investments. The narrative goes that AI will soon, and the operative word is "soon," replace all white-collar jobs, and with humanoid robots, replace all blue-collar jobs too. Maybe. Or maybe not. That's above my pay grade, but what I do know is that no one can predict if or when that will happen. So what do we do? We soldier on, taking on a little more uncertainty and taking to heart that, as William Gibson once said, "The future is here. It's just not evenly distributed." To confidently navigate buyout investments in the age of AI, we need to know exactly where the future is, on an investment-by-investment basis.

AI certainly provides rewards, but it also poses risks. It turns out the risk-reward calculus is two sides of the same coin, and the only way to solve for either is to understand what AI can do during the hold period. This takes us back to the Atomic AI Abilities first principles we were introduced to in Chapter 1. In this chapter we will apply our knowledge of AI abilities to assess risk.

Carrying on with the Gibson analogy, we need to know where AI can potentially disrupt a business—in other words, where the future is. To do so we use Figure 1 to guide us through our AI Impact Assessment of the different value propositions of the business, from the task level all the way up to the company level, to identify where the risks and, paradoxically, the rewards lie.

Figure 1 **AI Impact Assessment Template**

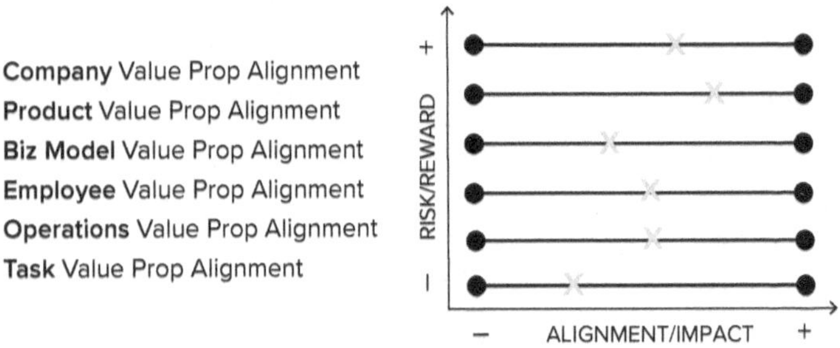

AI RISK IDENTIFICATION

There are of course other risks to consider (such as data quality, change management, execution and so on), but the risks considered here are AI investment risks, so our assessment considers the value proposition at each level of investment risk by asking, *How aligned is it with AI abilities?* Since AI will advance over time, this is a bit of a moving target, but generally speaking, the higher the alignment today, the greater the likelihood of risk at that level of value proposition.

If the value proposition of an important **task** within the business aligns with AI abilities, then a competitive company could use an AI to do that task. However, replacing a task, no matter how important, is the least of our competitive worries.

Next, if a costly or important **operation** is aligned with what AI is capable of, it's possible a competitor could use AI in that way. While more threatening than a task, a competitive company with greater operational efficiency is something, but it's not a huge risk to consider.

Next, if an **employee** value proposition is aligned with AI abilities, a competitive company could replace this type of employee with an AI. Depending on how reliant the business model of the company is on this type of employee, it may or may not produce a meaningful competitive advantage for an AI competitor.

Next up, if a potential new **business model** is aligned with AI abilities, a competitive company could disrupt the industry if the company were to deploy the AI.

Now we consider a **product**'s (service's, place's) value proposition. If what it does is aligned with AI's abilities, this would be an area of concern if a competitive company used AI to increase the competitiveness of a competing product.

Finally, we consider the value proposition of the **company** itself. If that value prop is in the wheelhouse of AI abilities, then a competing company using AI for this could be an existential threat. For example, if a large language model (LLM) could be trained to produce the same type of content as a company main value proposition, then there's a risk that the entire company could be disintermediated by an AI-enabled competitor.

This type of risk assessment is generally done during diligence to identify disintermediation risk. As we'll see in the next chapter, the same AI Impact Assessment is also used to identify the potential rewards from using AI for value creation. Risk and reward—again, two sides of the same coin.

During risk assessment, only those value propositions with high alignment identify potential risks, and the magnitude of these risks is dialed up or down based on the value prop level. Value prop alignment with AI at the company level is riskier than at the product level, which is riskier than at the business model level, and so on, down to the task level.

RISK ASSESSMENT EXAMPLE

Consider two companies, A and B, each with a different value prop alignment profile.

Company A (Figure 2) is potentially at risk if AI is not part of the value creation plan and is potentially rewarded if AI is part of the value creation plan. For example, a company recently seen in diligence specialized in producing highly technical documentation to support drug discovery and clinical trial submissions. The company's business model relied on teams of PhDs and regulatory experts compiling data packages, summaries and compliance documentation that met strict FDA standards. The firm walked away, acknowledging it wasn't prepared to embrace AI yet.

Figure 2 **Company A Risk Assessment**

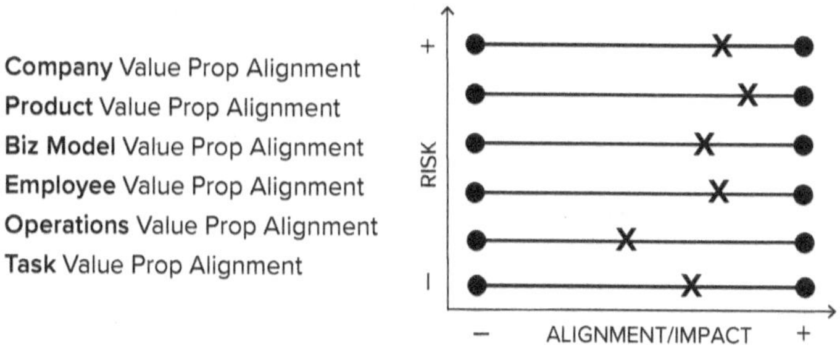

Company B (Figure 3) is potentially not at risk if AI is not part of the value creation plan and is not at risk or rewarded if AI is part of the value creation plan. A hypothetical counter example where AI would not be an existential risk would be a biotech firm focused on first-in-class drug discovery requiring deep scientific intuition and hands-on lab experimentation. While AI could aid in data mining tasks, the core value lay in creative molecular design and real-world bench science—tasks that demanded human involvement.

Figure 3 **Company B Risk Assessment**

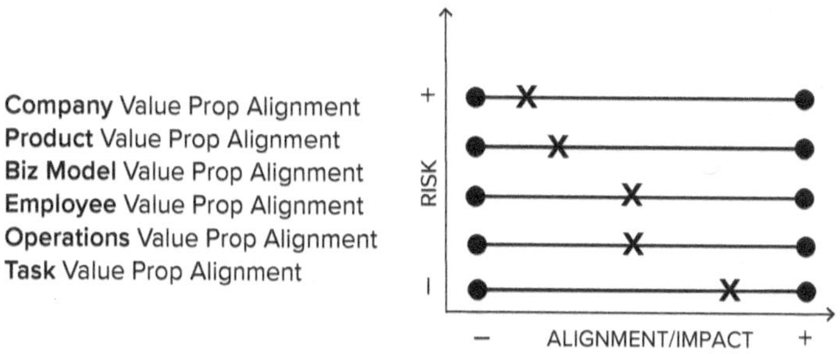

Risk then is in the eye of the beholder. The same assessment carried out to identify risk also identifies reward, or lack thereof—so it just depends on what side of the coin the sponsor lies.

SPONSOR RISK

Until now, we've been looking at the impact of AI at the company level, but over time the implications for PE firms will be just as, if not more, consequential.

An AI Impact Assessment identifies both opportunity and liability. If the sponsor invests in AI, it's an opportunity. If the sponsor is not ready to deploy AI as part of its value creation plan, then it's a potential liability.

If a PE firm buys a company like company A and does not equip it with AI, there's a risk that a competitive company, using AI, will outcompete its investment and negatively affect its value. If the PE firm doesn't buy the company, then it has lost out on a potential investment. Not entertaining the use of value-creating AI may introduce new deal risk and limit future deal flow. This is already happening and will only become more commonplace over time. The future is here.

CHAPTER 4

DEVELOPING AI STRATEGY

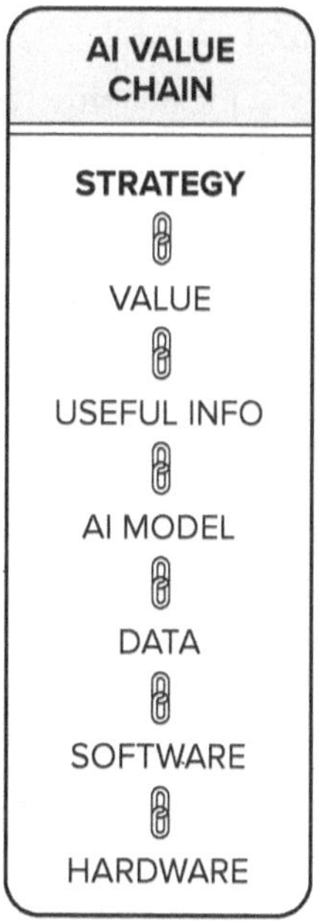

When you ask, what is my AI strategy, what you're really asking is, how can I best use AI to increase the EBITDA and valuation multiple of my company?

I suppose some companies could have a strategy other than increasing their value, but not PE-owned companies. For those companies their AI strategy drives or supports their investment thesis. This means strategy is at the top of the value chain and everything we do in this book is about increasing enterprise value in a way that achieves strategy.

Up to this point we have demonstrated how to create value with AI, not what value to create with AI—our strategy.

STRATEGY DEVELOPMENT

Developing a company's AI strategy pre- or post-acquisition differs in fidelity, but the high-level process of choosing what we're going to do with AI is the same. It always starts with how AI can improve business fundamentals such as growth plans, market trends, the competitive landscape and sponsor-specific goals. The result is an initial list of potential AI initiatives.

These initiatives will fall in one of two categories. Either they will augment the company's non-AI strategic efforts, or they will represent new strategic efforts that can only be realized with AI.

Ideally this work is started during diligence so the AI candidates can be prioritized among the other value creation candidates when developing the overall value creation plan.

This isn't to say we can't develop a company's AI strategy post-acquisition, because we can. It's just that depending on how far along the company is in executing its value creation plan, there may be less room to add new strategic initiatives. If that's the case, the focus will be on how AI can augment the existing value creation plans; but I think we can all agree that if an AI game-changer is identified, there's always room for a strategic course correction, especially when there is the possibility that the company's competition will also be using AI.

AI Impact

AI can impact business at many levels. When narrowing down the AI initiatives list of the items that make up our strategy, we look at the value proposition of each AI we are considering and ask ourselves, how aligned is it with the Atomic AI Abilities outlined in Chapter 1, and then, how rewarding (impactful) will it be?

The higher the alignment with AI abilities, the greater the likelihood of achieving or improving that value proposition. Using the same AI Impact Assessment we used in Chapter 3 for risk mitigation, let's look at each type of value proposition considered, starting with what is generally the least rewarding (tasks) to most rewarding (company), as shown in Figure 4.

Figure 4 **AI Impact Assessment Template**

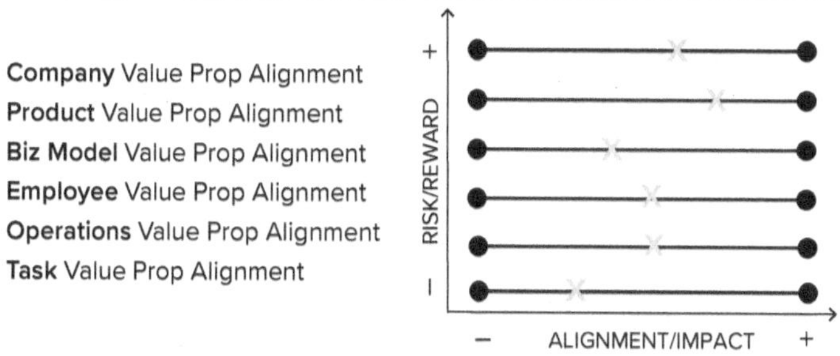

Starting with a task, if the value proposition of an important task aligns with the AI abilities, then AI can probably do that task instead of an employee. Although this is the least rewarding value proposition for the business, many companies stop here in their AI investigation.

Moving up, the more aligned a costly or important **operation** is with what AI is capable of, the more likely AI can improve the efficiency of that operation.

Next, if an **employee** value proposition is aligned with AI abilities, then we may be able to replace an employee or not hire another person to do the same job. The more aligned what the employee does is with what AI is able to do, the more likely an AI agent can do the job.

Moving up further, we consider how aligned a new **business model** is with AI abilities. If it is, the potential reward can be game-changing.

If what a company's **product** (service, place) does or wants to do is aligned with AI's abilities, then we can likely innovate the product with AI to make it more competitive.

At the top of the value proposition list is **company**. If the strategy of the company, at a high level, nicely aligns with AI abilities, then it's a good indicator that alpha AI value can be created in this company.

Only those value propositions with high alignment are considered part of the AI strategy, further whittling down the candidate list (Figure 5). Generally, replacing a task with AI will be less impactful on a company's valuation than replacing an operation. And innovating a product to grow revenue will generally be more impactful than lowering operational or employee costs. From doing an AI Impact Assessment for value, initiatives that plot to the right are more likely to be part of an AI strategy than those that plot to the left. Once the list of strategic initiatives has been sufficiently narrowed, we can then do an "AI readiness" or "capabilities gap" analysis to get a rough idea of what effort will be involved.

Figure 5 **AI Impact Assessment**

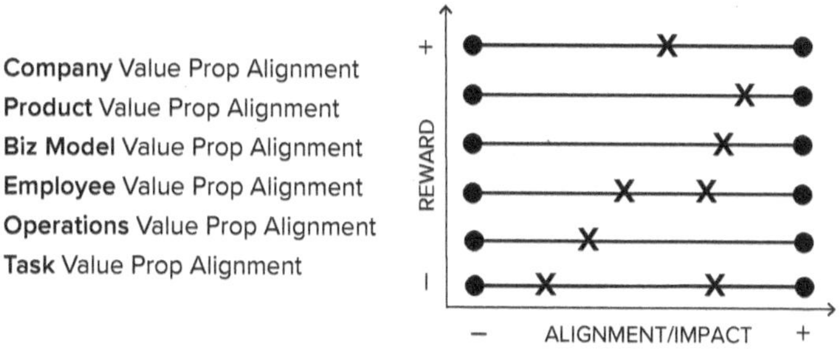

At the end of the day, the tangible output of an AI strategy is a list of strategic initiatives, each of which is realized with one or more AI apps or agents. This means that a company's AI strategy consists of one or more AI applications or agents to be integrated into the company's products or operations.

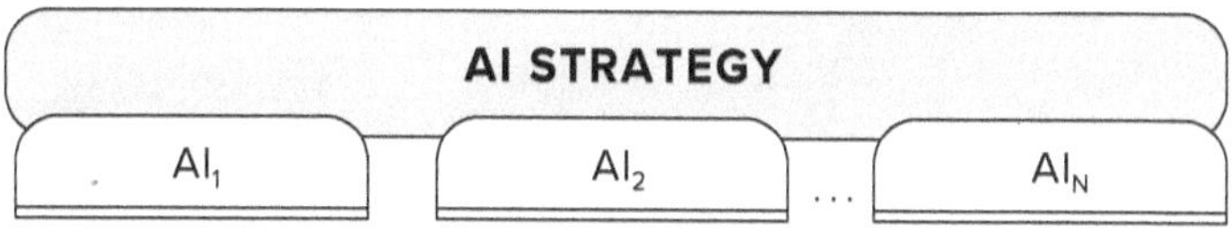

AI STRATEGY = one or more AI Applications or Agents

• • •

Given that, it's probably a good idea to get into what AI applications and agents are. We do that in Chapter 5.

CHAPTER 5

AI APPLICATIONS AND AGENTS

AI applications and agents are the holders of the AI model. In fact, you can think about the AI app or agent as the wrapper that goes around the model so it can interface and operate with or within a product (service, place) or operation to create the value.

AI applications are software that uses one or more AI models to achieve preset goals. They're used in structured and static environments to solve a specific problem (Figure 6).

In contrast, AI agents are software that cycles through PERCEIVE AI models, REASON AI models and ACT AI models and back again to pursue goals autonomously, and they're used in dynamic and goal-driven environments where autonomy and adaptation are necessary (also Figure 6).

For example, in logistics and shipping, an AI application could be a tool that uses Geospatial AI for route optimization, suggesting the shortest path for deliveries, whereas an AI agent would act as a logistics teammate that continuously reroutes trucks in real time based on traffic, weather, and order priority, while also updating customers on expected

Figure 6 **Comparing AI Application and AI Agent Software**

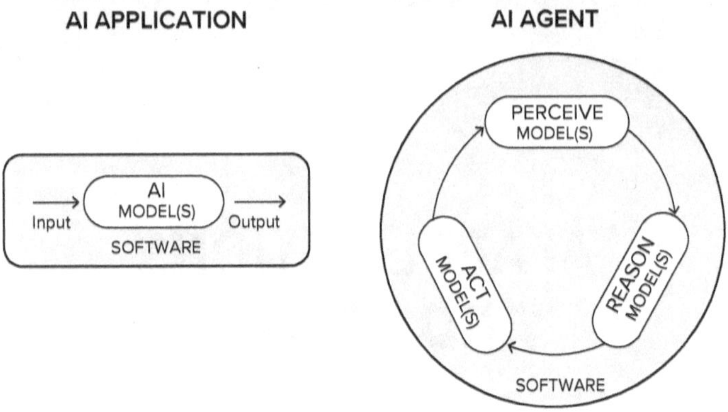

delivery times and renegotiating supplier deadlines. This is pretty lofty stuff, but it is commercializing fast. In the next part of the book, I'll showcase one of my clients that is using AI agents to augment or replace marketing employees.

It should also be noted that hierarchies of AIs are frequently used. Usually, orchestration code is used to manage multiple AIs. And when this orchestration layer manages AI agents, it can be called an agentic workforce. Agentic. You'll be hearing that word a lot.

In this book when we say AIs, we are not talking about AI point products like GenAI chatbots, at least not through a prompt window—we mean AI applications and AI agents (see Figure 7). AI point products are terrific for helping with point tasks and getting familiar with this specific type of AI, but as we've seen from the AI Impact Assessment, a chatbot-quickened task is useful but generally not part of any meaningful AI value creation strategy. Using the same off-the-shelf AI point products (such as Microsoft Excel) that everyone else does for, say, software development, productivity and go-to-market motions can be very useful, but they don't meaningfully move the valuation needle.

Figure 7 **Different Types of AI**

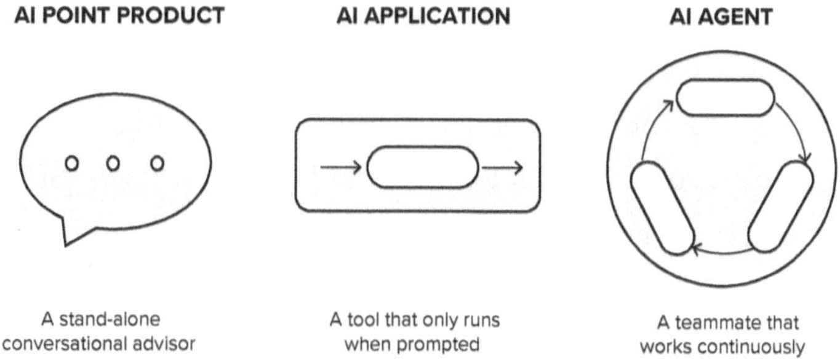

Having said that, we always use off-the-shelf AI apps and agents, and generally, but not always, we interface with these AIs through their application programming interfaces (APIs) to integrate them into our products or into the workflows of company operations.

While off-the-shelf products are used, the AI IP is absolutely developed. This always involves capturing the tuning parameters for the AI models we use. The IP also includes the training methodology to discover the parameters and how the AIs are integrated into the business. This IP should always be transferred to the company; and down the road, a small internal team should be hired to advance it.

While it can't tell you exactly which AI applications and agents to include in your AI strategy, this book provides the playbook for how to identify them, define them, quantify them and build them.

THE DIGITAL TWIN

In the context of value creation, a digital twin is a virtualization of how a physical product creates value.. It exists as an AI model wrapped in software. Once in this digital form, the physical product is software-defined and data-driven, enabling it to benefit from AI just like a software product can. Using the steam boiler as an example, its digital twin doesn't represent how it looks; instead, it represents its value: how natural gas is burned to heat up steam (Figure 8). As will be discussed further in an upcoming case study, once we modeled the combustion process mathematically, it was then used to improve boiler efficiency.

Figure 8 **The Digital Twin of Steam Boiler Combustion**

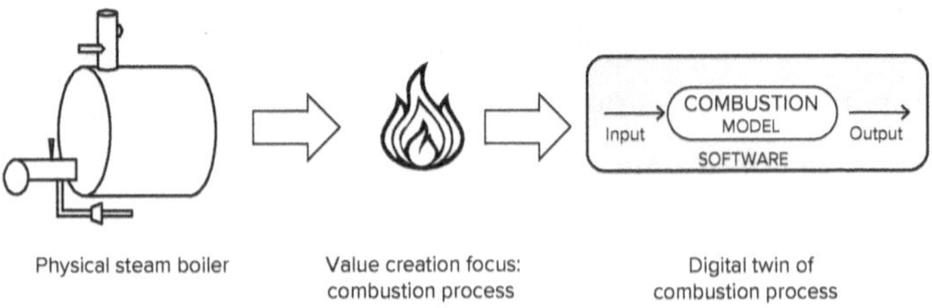

SYSTEM INTEGRATION

For AI to be part of a business, it needs to be part of its business systems. At the task level, AI point products are generally stand-alone and integrated into workflows by hand. But when used as apps or agents, the useful information outputted from the AIs is integrated into the company's products and operations.

Product integration is bespoke. The software product or the physical product's digital twin integrates with the AI software (Figure 9). Either way the AI is in the form of a third-party API or a library or as custom raw code.

Figure 9 **Integrating AI Applications and AI Agents**

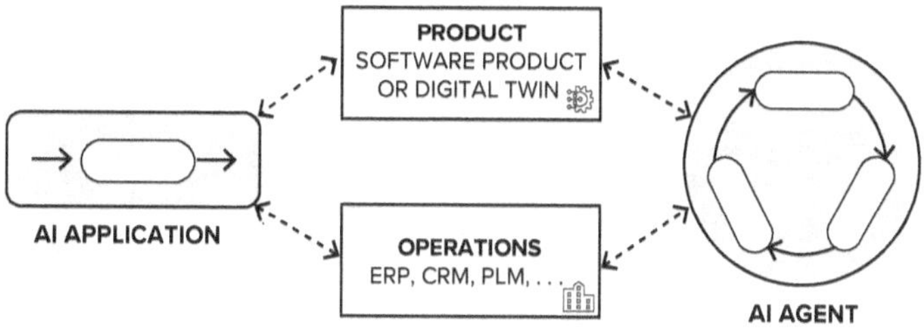

Operations integration connects the system of record with the AI software. Whether the system of record is an ERP (Enterprise Resource Planning), CRM (Customer Relationship Management), PLM (Product Lifecycle Management) or another business system depends on the value being created.

Note: **For the rest of this book, AI applications and agents will simply be called AIs, and a single AI application or agent will be called an AI.**

In Part One we showed that your AI strategy consists of one or more AIs that possess one or more AI abilities. As an example, an AI strategy could be focused on predictive maintenance, where the AIs must first DETECT ANOMALIES in IoT data by flagging out-of-pattern signals indicating potential failures (PERCEIVE), then INTERPRET what these anomalies mean by correlating anomalies with previous failures to understand causes (REASON) and finally PREDICT when things are going to go wrong by forecasting time-to-failure to schedule preventative maintenance (ACT).

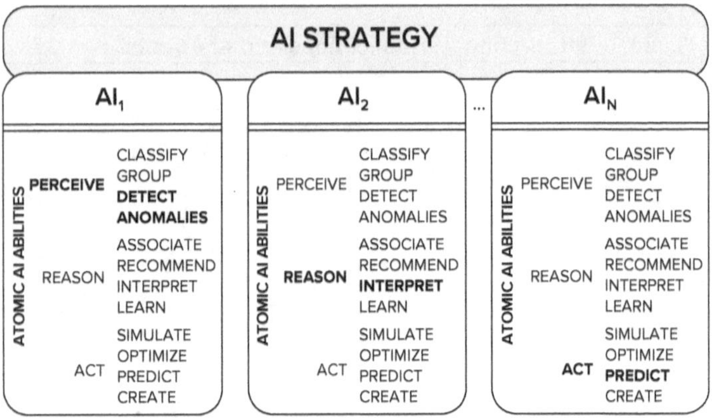

Once one or more AIs have been identified to carry out the AI strategy, the next step, or concurrent step, depending on your workflow, is to further filter these AI candidates by quantifying the value they create financially. Part Two of this book enables you to do this—to rank-prioritize the identified AIs based on how much value they create.

• • •

To reinforce this novel way of thinking, we'll show how to apply the techniques learned here in Part One by demonstrating their use in 10 value creation case studies in Part Two.

PART TWO

VALUE DRIVERS

Part One of this book covered how to create value with AI. In Part Two we'll discuss the types of value that we can create.

Value type is first grouped by the value drivers—revenue growth, cost reduction and multiple expansion—and then further classified by the value levers: market share expansion, customer expansion, product expansion, market expansion, customer retention, pricing optimization, margin improvement, operational efficiency, add-on acquisitions and the multiple rerate. Once value is classified this way, we can quantify the value produced, as we'll demonstrate in Chapter 6.

AI VALUE CHAIN

STRATEGY
VALUE
USEFUL INFO
AI MODEL
DATA
SOFTWARE
HARDWARE

To apply each of the value drivers, we "pull" different value levers:

Revenue Growth	Cost Reduction	Multiple Expansion
Market Share Expansion	Margin Improvement	Add-On Acquisitions
Customer Expansion	Operational Efficiency	Multiple Rerate
Product Expansion		
Market Expansion		
Customer Retention		
Pricing Optimization		

These value drivers and their value levers are not unique to AI value creation. In fact, every form of value creation and operational improvement can be classified more or less the same way.

Having a common classification system is key to being able to compare the efficacy of AI value creation with other forms of value creation. While this book is about using AI in value creation, in practice AI will be only one of a number of value creation options available. Quantifying them all the same way enables us to invest in those value creation initiatives with the highest ROI/ROCE—independent of whether they are AI-based or not.

• • •

After the "big picture," value drivers and levers are introduced in Chapter 6, the following three chapters will cover the value drivers: revenue growth, cost reduction and multiple expansion.

CHAPTER 6

QUANTIFYING AI VALUE

Each AI we develop is aligned with our strategy and drives the worth of our company up through revenue growth, cost reduction or multiple expansion.

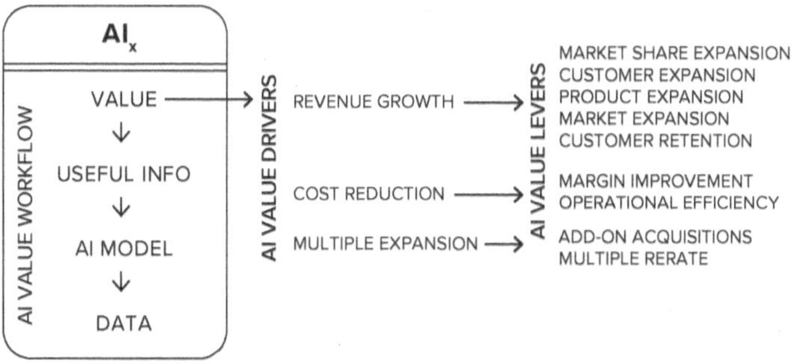

Once classified by value driver and lever, it's relatively straightforward to quantify an AI's impact on enterprise value.

My view is if you can't quantify its value, your AI project is better described as a hobby or experiment rather than an investment.

FINANCIAL IMPACT

To demonstrate an AI's financial impact, let's use an example company with the profit and loss (P&L) metrics from about 1,000 midsize ($100 million to $1 billion) US public companies (McKinsey, 2017), normalized to a baseline index of 100 and with a valuation multiple of 11, to compare the impact of each value driver. Furthermore, we'll use profit for EBITDA, cost of goods and services for variable cost, and fixed cost as the difference between EBITDA and gross profit (left side of Figure 10).

Figure 10 **Impact on Enterprise Value**

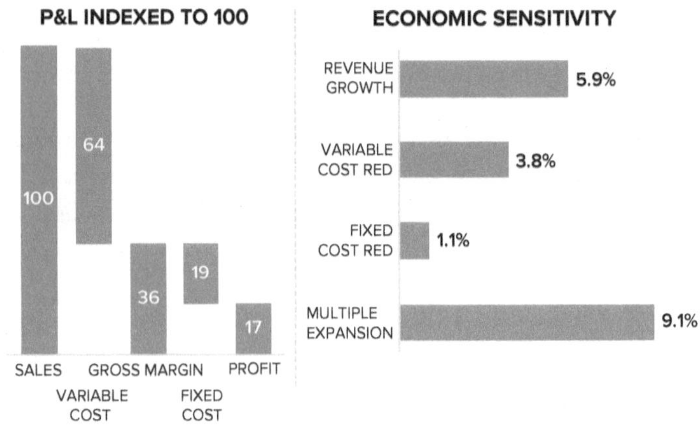

Economic sensitivity analysis (right side of Figure 10) demonstrates:

- A revenue growth of 1% increases the company's enterprise value by 5.9%.

- A variable cost reduction of 1% increases the company's enterprise value by 3.8%.
- A fixed cost reduction of 1% increases the company's enterprise value by 1.1%.
- A multiple expansion of 1 increases the company's enterprise value by 9.1%.

Assuming similar costs, this analysis shows multiple expansion and revenue growth have the highest impact—something that should be considered while formulating your AI strategy.

Substituting in your company's P&L metrics along with value driver assumptions allows you to financially model AI impact on enterprise value.

Value Bridge

Although the value levers are interrelated and affect each other, we are going to largely ignore their positive compounding effects and instead address each driver individually, independent of its effect on the others.

The value bridge diagram (Figure 11) shows the five lifts that contribute to an increase in enterprise value, as summed by the last column, "sell." Note that the "buy" and "industry" lifts are excluded from the AI alpha.

This book doesn't cover how to produce the buy lift (column 1 in Figure 11) that comes from financially engineering the deal structure or leverage. Nor does it cover the industry

Figure 11 **Value Driver Alpha**

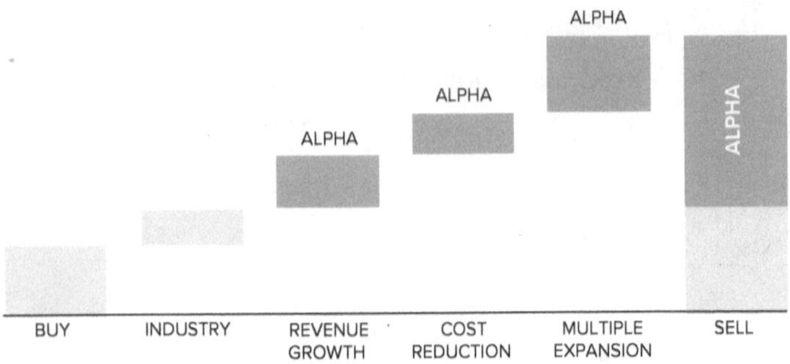

lift (column 2) that naturally occurs over time by choosing the right company in a market with positive dynamics.

In this book we're focused on the AI efforts that result in alpha lifts in revenue growth, cost reduction and multiple expansion.

AI VALUE LEVERS

Before we dive into each of the value levers in the next three chapters, let's look at them in aggregate.

Revenue Growth

1. **Market Share Expansion:** Grow revenue by selling products (services, places) to new customers in current market.
2. **Customer Expansion:** Grow revenue by selling more products to existing customers.
3. **Product Expansion:** Grow revenue by selling new products to existing and new customers.

4. **Market Expansion:** Grow revenue by selling products to new customers in new markets.
5. **Customer Retention:** Grow revenue by keeping existing customers.
6. **Pricing Optimization:** Grow revenue by selling products for their true value.

Cost Reduction

7. **Margin Improvement:** Reduce costs by increasing profit margins.
8. **Operational Efficiency:** Reduce costs by optimizing operations.

Multiple Expansion

9. **Add-on Acquisitions:** Expand the valuation multiple by growing inorganically.
10. **Multiple Rerate:** Expand the valuation multiple by earning a higher comp.

• • •

Next, we'll dive into each of the 10 value levers in more detail. Chapter 7 covers the six value levers that grow revenue. Chapter 8 covers the two value levers that reduce costs. And Chapter 9 covers the two value levers that expand the valuation multiple.

In each of the upcoming chapters I will outline the value levers and share a case study for each, all of which come from work I have done for past clients. This will provide deeper learning for each lever and will further reinforce what was learned in Part One of this book: how to create value with the AI value creation workflow and framework.

CHAPTER 7

REVENUE GROWTH OPPORTUNITIES

Revenue growth means increasing a company's top-line sales. It directly influences valuation by driving EBITDA higher and lowering perceived risk.

This chapter is about using AI to acquire new customers in existing markets (market share expansion), selling more products (services, places) to existing customers (customer expansion), launching new products (product expansion), entering new markets (market expansion), reducing churn (customer retention) and better aligning pricing to value (pricing optimization) (Figure 12).

Figure 12 **Revenue Growth Value Levers**

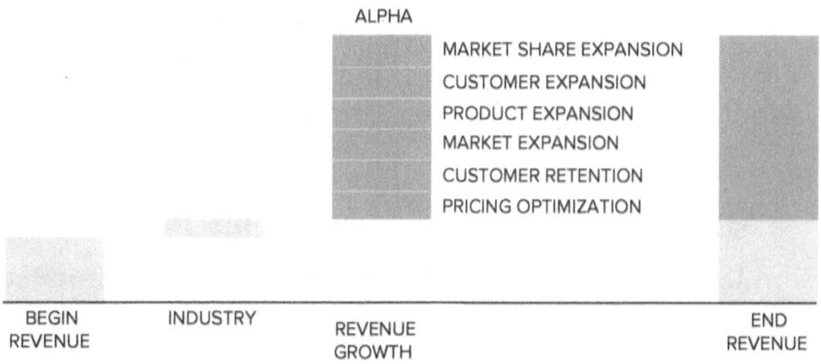

Customer Sims

Although we can use AI in a variety of ways to boost revenue growth value levers, a common thread for each is to use AI to better understand the customer. And by better understanding the customer, I don't mean understanding the ideal customer profile (ICP); I mean understanding each customer individually.

Customer sims (simulations) are AI-based simulations of customer behaviors and product interactions, used to optimize sales and marketing (Figure 13). Two novel product interaction examples are usability and utility models.

Figure 13 **Customer Sim AI Model Types**

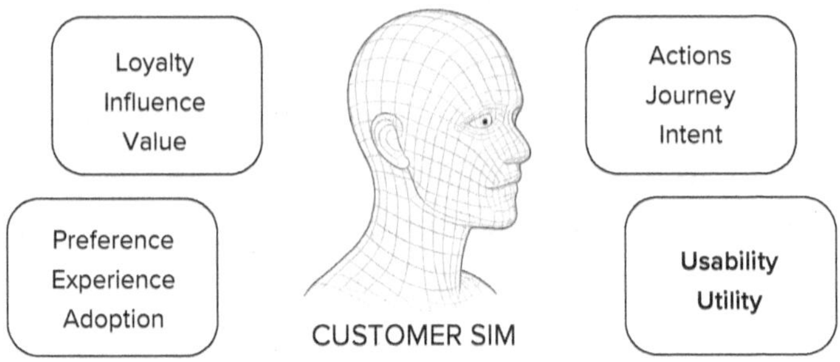

A customer sim usability model is a simulation of how the customer uses the product. Heat mapping is common in e-commerce, but now with the advent of the digital twin, usability mapping can be used for any product—digital or physical.

Same thing for a customer sim utility model. It is a simulation of what the customer does with the product.

Considered sophisticated in the past, with the increased accessibility of AI tools and the digital twin, it too can be applied to any type of product—digital or physical.

Customer sims have an outsized role in sales- and marketing-focused value creation. Through these more sophisticated models, AI allows us to understand the customer better than before.

System Integration

As discussed in Chapter 5, no AI can be an island. To meaningfully produce value, it must be integrated with the corresponding system of record. In this case of revenue growth, AI's useful information is primarily networked to the company's CRM and PLM systems to be put into action and often to connect to support ticketing and marketing automation systems.

All right, now that we've addressed sims and integration, let's go through each revenue growth lever, one at a time.

MARKET SHARE EXPANSION

Market share expansion involves growing revenue by selling more of a company's existing products or services to new customers within current markets. It is a key growth lever because it lowers risk by demonstrating scalability.

Traditional Approaches

Traditional market expansion relies on traditional marketing and funnel management, intuition-driven promotions and sales force initiatives.

How AI Can Help

AI market share expansion helps by boosting existing marketing efforts, using proprietary customer data to individualize marketing, and by sparking product innovation (my favorite).

Boosting Traditional Marketing and Sales

Almost every aspect of marketing and sales can be improved with AI models and automation. Customer sims give us a higher-fidelity view of the customer, resulting in more effective traditional marketing and sales efforts.

To level up, customer sims can be combined with AI employee agents to simulate customer-company interactions. For instance, lead scoring AI agents that I've worked with (see the margin improvement case study) have proved to be more effective than their human counterparts.

Data-Driven Marketing

Whereas traditional marketing addresses segmented cohorts, data-driven marketing uses continuously flowing customer data to personalize communications, addressing each customer individually. This high-fidelity customer data has been, until now, only available to properly architected software products; but with the digital twin, fed by the internet of things, all products and services can be software-defined and data-driven, enabling them to use these techniques. And AI-driven analytics can optimize sales territories to improve sales productivity and capture market share more effectively.

Product Innovation

Using AI, we can make our products more competitive by improving their value propositions with new features, higher performance and better experiences.

By using a product sim to simulate how the product creates value, we can optimize it for performance, efficiency or any other product key performance indicator (KPI) important to the customer.

Client Case Study: Building a Better Mousetrap

It's fitting that my first case study is from my very first consulting client. A family-owned consumer and business pest control trap OEM wanted to expand its presence in the highly regulated commercial food production (baking) market, so it looked to innovate its 50-year-old product.

We used an AI analytics system to identify the best trap locations (system design) and an IoT system to reduce the manual checking of traps. We did this by instrumenting the traps with sensors and inputting their proprietary trap data (number of mice caught per trap, where, when and so on) to tune an AI model that continuously optimized the trap network by moving and eliminating nodes (traps). Not only did automated trap checking reduce the time-consuming and costly “walkarounds” (checking traps manually), but it effortlessly created the pest control service logs required to meet food production regulatory compliance.

The more innovative smart trap enabled my client’s third-generation owners to expand their business by moving into the digital age.

Value Creation Framework

MARKET SHARE EXPANSION AI

VALUE
End customer: Lower CAPEX by reducing the number of traps, and lower OPEX by automating trap monitoring and paperwork generation.

↓

USEFUL INFO
Location for each trap and regulatory information on trap monitoring and bait placements.

↓

AI MODEL
SIMULATE → OPTIMIZE
SIMULATE the “network” of traps; then OPTIMIZE the network by removing and moving “nodes” (traps).

↓

DATA
Current trap locations (map), interior with negative and positive spaces (map), trap activity (IoT).

Note: **For this case study and the rest that follow, the data in the AI Value Creation Framework is for inferencing, not training, and the sources of data are parenthesized.**

CUSTOMER EXPANSION

Customer expansion involves growing revenue by selling more products or services to existing customers, thereby increasing customer lifetime value. It is a key growth lever because it boosts revenue and profitability without incurring high customer acquisition costs.

Traditional Approaches

Traditional customer expansion relies on manual cross-selling, upselling and loyalty programs, which depend heavily on instinct.

How AI Can Help

AI customer expansion helps by analytically identifying customer needs and personalizing the upselling and cross-selling interactions with them and then networking the useful information to existing digital infrastructure such as CRM, marketing automation and e-commerce platforms.

Identifying Customer Needs

The first step in AI customer expansion is to use customer and product sims to analyze customer behaviors and purchase histories to pinpoint upsell and cross-sell opportunities.

Personalized Upselling

"Land and expand" is the upselling technique that begins by offering a fully loaded digital or physical product for free or a low price and then upselling new features when more of the product is needed. This has been done before with software. What's different now is this technique can be applied to all types of products, with AI choosing what feature to sell, to which customers, when.

Mainstream examples include Adobe upselling to a paid PDF Acrobat when editing is needed and Tesla upselling a longer battery range by flipping a switch. And a past client upsold safety features during an incandescent to LED conversion program for campuses (see the market expansion case study).

Personalized Cross-Selling

Personalized cross-selling takes two approaches: the car warning light approach and the Netflix recommendation approach.

When one product depends on a second product like an ICE car needs oil or a razor needs blades, the manufacturer is in the best position to cross-sell the secondary product. The cost of sale here is the software and AI that decides the best time to alert the end customer when the secondary product is needed. My HP printer does this to cross-sell me ink cartridges, and a past industrial client of die cutting machines did this to upsell cutting fluids and blades.

The Netflix approach uses AI to recommend new products (movies and television shows) to the end customer based on potential needs (watching habits). For SaaS (Software as a Service) products it's about increasing consumption to signal value. For physical products my favorite is to cross-sell fintech products. By using customer and product sims, the OEM knows enough about the customer to always underwrite the product (insurance, financing and so on) at a profit. Tesla's auto insurance is an example of this, as is the following case study.

Client Case Study: Underwriting Extended Warranties

A 60-year-old family-owned boiler manufacturer had a large install base with very few customers under warranty. We decided to invent a high-margin personalized product to cross-sell into the company's install base.

We used a stand-alone AI system that ingested the clients' proprietary data to output those install base customers who would be profitable if they purchased an extended warranty for their boilers. We did this by correlating IoT boiler data (see the pricing optimization case study) with install base data (boiler type, industry, use, age, service history and so on) to tune an AI model to identify which customers to sell the low-cost but profitable extended warranty to.

AI and IoT enable every OEM to cross-sell profitable fintech products to increase EBITDA.

Value Creation Framework

CUSTOMER EXPANSION AI

VALUE
End customer: Protection against unexpected repair costs ensuring continued operations at lower operating costs.

↓

USEFUL INFO
List of extended warranty customer candidates who have boilers that are less likely to require warranty repairs.

↓

AI MODEL
CLASSIFY → LEARN
CLASSIFY customers into different boiler categories; then LEARN which categories need the least suppor.t

↓

DATA
Boiler data (IoT), install base data (business system).

PRODUCT EXPANSION

Product expansion involves growing revenue by introducing new products or services to existing customers and new customer segments. It is a key growth lever because it enhances revenue and reduces product concentration.

Traditional Approaches

Traditional product expansion relies on small-sample customer surveys, market research, competitor benchmarking and internal product development ideas driven by intuition.

How AI Can Help

AI product expansion helps by enabling data-driven product invention and using proprietary customer data to fine-tune the selling of those new products.

Product Invention

Understanding customer needs is simple in concept but difficult in practice. Fortunately, by looking at the continuous data coming from the digital product or the physical product's digital twin, we have a 24/7 view into what the customer does with our product.

To increase the probability of success beyond intuition, AI analyzes usage patterns and compares the results with the customer's most important KPIs to identify new product opportunities.

Another path to AI invention is to identify what the customer does with the product via usability models differently from what the product was designed to do. This provides signals to direct product invention.

Data-Driven Sales

Whether selling existing products or new products, we use AI to go beyond product-market fit. To help us find the right customers, we use customer sims and other AI to group buying customers for each product, resulting in product-segment fit. This can be further refined by using individual customer data to find product-customer fit. These data-driven sales techniques equip salespeople with the useful information they need to close sales.

Client Case Study: Improving Hip Replacements

A surgeon-owned medical device company wanted to expand its hip replacement product line by inventing a new product to gain traction in this highly competitive market.

Improper fit between the bone and the artificial hip joint is a leading cause of patient pain and redress surgeries. We developed an AI-based reaming tool motor control software that prevented necrosis (bone cell death) caused when the surgeon drills the hemispherical hole in the patient's pelvis to hold the artificial hip ball joint. The fewer dead cells, the better the fit. We did this by developing an AI model to simulate necrosis. Then we reorganized the model's equation to control the speed of the rotating reamer head so that the reamer never raised the temperature of the bone cells high enough to lead to necrosis (about 113°F). No matter how hard pressed or how long used. The reamer's rotational speed would slow to ensure the heat of the drilled bone cells never surpassed the temperature that would kill them.

This product direction expanded my client's IP, resulting in a joint venture that ultimately led to the company being acquired.

Value Creation Framework

PRODUCT EXPANSION AI
VALUE End customer: Shorten hip replacement surgery time and reduce follow-on redress surgeries. ↓ USEFUL INFO The correct rotational speed of the reamer head under all device conditions. ↓ AI MODEL PREDICT PREDICT bone temperature; then pass to the app to determine the new rotational speed. ↓ DATA Current rotational speed (IoT), pressure (IoT), time (app).

MARKET EXPANSION

Market expansion involves growing revenue by selling a company's existing products or services to new customer segments by entering new market adjacencies. It is a key growth lever because it reduces dependency on existing markets.

Traditional Approaches

Traditional market expansion relies on manual market research and surveys, spreadsheets, competitive benchmarking and gut-based customer segmentation.

How AI Can Help

AI market expansion helps by analyzing generally available data such as competitor presence, customer spending, market demographics, market saturation and pricing structures to find patterns that point to potential adjacent markets. What was once only possible with a full data science team can now be done efficiently by almost any STEM-minded person with commercially available AI tools.

Seeing Things

Where things get more interesting is to "intersect" this public data with a company's proprietary data to find adjacencies that are there but invisible to the eye. Due to the multidimensional data involved, AI analytics is better suited to find customer insights by detecting cluster anomalies and outliers in the market expansion data.

Proprietary Adjacency Data

Discovering customer commonalities is impossible without collecting the right data. New adjacencies are found by examining vertical integration (expanding across the value chain) and horizontal integration (expanding across similar markets or competitors) data. Beyond this data we use AI analytics to find patterns in other data we directly or indirectly capture from the customer. This includes proprietary data on customer geographies, verticals, use cases, distribution, company size and customer demographics (see Table 2).

Table 2 **Proprietary Market Adjacencies Data**

SEGMENT	DATA TO COLLECT ON EACH CUSTOMER
Geographic	Country, city, zip code
Vertical	Industry, sector, market segment
Use case	Product use
Distribution	Channel, partners
Company size	Small or midsize to global enterprise
Customer	Age, income, education

Generally, companies are not collecting all this data on each customer. AI market expansion requires that this data be researched for existing customers and captured for new customers. Capturing this data for new customers may require new tech and incentives for sales and marketing employees.

Client Case Study: Streetlight Conversion

A private e-commerce company selling illumination products to schools landed a large incandescent-to-LED conversion contract. The company wanted to vertically integrate its business into school campuses by becoming an OEM of a smart lighting product line.

After identifying safety as the key value proposition, we developed a new streetlight head that could be aimed, focused and brightened. We did this by using an AI system that ingested the location of lampposts on a map and output the direction, focus and brightness for each lamppost's streetlight head to illuminate the campus with consistent luminance values 24 hours a day.

The contract was a vehicle for my client to expand into an adjacent market: customers willing to pay a higher price for the additional value proposition (improved campus safety) and make structural changes that would eventually result in a higher valuation multiple.

Value Creation Framework

MARKET EXPANSION AI

VALUE
End customer: Improve campus safety by ensuring all grounds are properly lit.
↓
USEFUL INFO
Direction, focus and brightness for each streetlight head.
↓
AI MODEL
CREATE → OPTIMIZE
CREATE inputs for illumination equations; then OPTIMIZE underconstrained equations.
↓
DATA
Streetlight (map), campus geography (map), time of day (app).

CUSTOMER RETENTION

Customer retention involves growing revenue by maintaining and strengthening relationships with current customers to minimize attrition and maximize repeat business. It is a key growth lever because lower customer acquisition costs lead to higher profitability, and higher retention rates lead to more predictable revenue, reducing risk.

Traditional Approaches

Traditional customer retention relies on periodic customer satisfaction surveys, loyalty programs, sales outreach and continuous customer support. They are often reactive and treat all customers equally rather than focusing on high-value accounts.

How AI Can Help

AI customer retention helps by predicting churn risks and then proactively addressing them by directing timely support interventions to maintain satisfaction and loyalty.

Early Churn Prediction

Even when traditional approaches examine customer data to identify churn indicators, it is often too late. Too late because they look at the past to see what happened.

AI is different. It looks at the present. It looks at real-time customer data and compares it with past data to make predictions about the future. It is this instantaneous analysis of purchase history, usage, behavior patterns and engagement data that allows for proactive intervention before customers decide to leave. And not only does AI customer retention identify flight-risk customers, but it can also prioritize the right customers to retain—the ones that buy more and cost less to support.

Churn Sim

This is done with a customer sim churn model that analyzes data about past customers who churned. These customer sims allow us to rank-stack every customer by how at risk they are and the difference between how much they're worth and how much they cost us to keep.

Automated Triage

Immediately identifying risk is phase 1 of an AI customer retention solution. Next, the risk must be rapidly triaged and addressed. The AI triage model is applied to about-to-churn customers we want to keep. The steps taken are:

1. Classify the type of risk by pattern matching to previous risks that were turned around.
2. Choose the best solution from past solutions that successfully addressed this type of risk before.
3. Autonomously generate and message customers to set the stage for resolution.
4. Dispatch custom solutions to the right person on the customer success team so they can personally engage with the customer to reach resolution.

AI helps customer success teams intervene more effectively, but we can't forget that human judgment remains critical.

Client Case Study: Smart Inventory Control

A sponsor-owned OEM of midstream gas compression equipment wanted to maintain and grow its higher-margin replacement parts and services business. However, in the gas-rich Permian Basin, if a customer's compressor went down and the needed replacement parts weren't available from the OEM's nearby shops, the customer would buy non-OEM parts from a closer competitor's shop so their compression station could be up and running again in a shorter amount of time.

To minimize customer ship time and increase retention rates, we replaced a decentralized inventory system of paper ledgers and databases with an AI system that informed supply decisions to route the right parts to the right shops. To reduce churn, we tuned the models of an off-the-shelf AI-based inventory control system with the historic data needed to predict when customers would need replacement parts. To get ahead of the predicted problem, these parts were shipped to the closest shop to the customer before the predicted need.

This reduced churn and provided our client with roll-up guidance for this platform company (see the add-on acquisitions case study).

Value Creation Framework

CUSTOMER RETENTION AI

VALUE
End customer: Shorten gas flow interruptions by replacing damaged parts in the shortest time.

↓

USEFUL INFO
An inventory topology that identified which parts should be in which repair shop and when they should be there.

↓

AI MODEL
INTERPRET → PREDICT
INTERPRET service logs; then PREDICT parts each customer will need, and use the app to identify the closest shop.

↓

DATA
Customer service history (business system), location of each customer (map), location of each shop (map).

PRICING OPTIMIZATION

Pricing optimization involves growing revenue by adjusting prices to reflect the true value of products, or at least the price the market will bear. It is a key growth lever because it ensures maximum profitability per transaction.

Traditional Approaches

Traditional pricing optimization relies on cost-plus pricing, manual competitor benchmarking, third-party customer surveys and interviews. Without proper justification, price increases risk customer alienation and reputational risk.

How AI Can Help

AI pricing optimization helps by interpreting proprietary customer data to find the right value-based price and then wrapping that price in the right business model.

Data-Driven Pricing

Data-driven pricing is pricing power. The power to set prices commensurate with the value created is the holy grail of pricing. Everyone knows this, but without using proprietary customer data it's impossible to do.

Since pricing is an exchange of value, we must quantify how much value/money our product generates for the customer. In AI pricing optimization, we identify the KPI that contributes to the end customer's revenue and then proactively capture the customer data we need to quantify that KPI.

Through financial projections, data-driven pricing then determines how much our product contributes to that KPI and prices it accordingly. Example KPIs are, per user, per usage, per transaction, per consumption, per job and so on.

AI pricing can be made transparent. When customers see a clear, data-driven reason for a price increase, they're more likely to accept it, especially when the company can demonstrate that the product remains cost-effective at the new price.

Data-Driven Business Models

In business model design the primary goal is to minimize monetization friction by aligning the product's business model to the right KPI in the customer's business model.

True value-based pricing is only possible with data-driven business models. Guessing this relationship without proprietary customer data won't be accurate or dynamic. Data-driven pricing models are available in every industry where we can quantify value exchange. A digital example is AI adjusting subscription pricing for an SaaS platform based on user engagement. This seems intuitive because we may have experienced it ourselves with software, but data-driven pricing is also applicable to physical products.

For example, a mining company's bucket wheel excavator's value is based on the amount of earth it excavates—the more it excavates, the more money it makes the mining company. How it makes money, and therefore its business model KPI, is per weight of earth removed. A one-time price for the excavator is not aligned with how its customer makes money. Another option is to price the product as a service (PaaS), often using a time-based business model—ensuring that the monthly fee is lower than the amount of money the product earns. Even more aligned with the customer's business model is to be paid by outcomes. Pricing the excavator with an outcome-based business model would be dollars per pound of earth removed. To enable the last two business models, we must add sensors to the excavator to measure the weight of the earth it excavates. The following sequence provides an alignment structure for business model design:

One-time price → PaaS → Outcome pricing

Note: If you're interested in data-driven business model design, my first book, *IoT Inc.*, published by McGraw-Hill, covers this topic in detail.

The following case study provides another example of business model alignment.

Client Case Study: New Boiler Business Model

The same 60-year-old family-owned boiler manufacturer from the customer expansion case study wanted to build an AI product to reduce the consumption of natural gas—the largest variable cost of generating steam.

We developed an AI system designed to improve the efficiency of boilers in their install base and the install base of their competitors by improving gas combustion. We did this by developing a custom AI model for burner tuning that was integrated into the IoT middleware we licensed. The proprietary data collected enabled a data-driven business model based on fuel cost as the KPI.

This allowed my client to offer its PaaS, initially based on a monthly fee priced less than the natural gas cost savings. This optimized the price to be better aligned with the product's value.

Value Creation Framework

PRICING OPTIMIZATION AI
VALUE End customer: Spend less on natural gas. ↓ USEFUL INFO Customized burner parameters to better tune the boiler. ↓ AI MODEL OPTIMIZE OPTIMIZE the underconstrained fuel-O_2 burner equation. ↓ DATA Fuel flow (meter), damper angle (IoT), O_2 (IoT).

CHAPTER 8

COST REDUCTION OPPORTUNITIES

Cost reduction involves strategically decreasing expenses to enhance profitability. It directly influences valuation by boosting EBITDA and by paying down debt.

This chapter focuses on using AI to improve profit margins (margin improvement) and to increase productivity (operational efficiency) by driving greater value from existing resources and operations (Figure 14).

Figure 14 **Cost Reduction Value Levers**

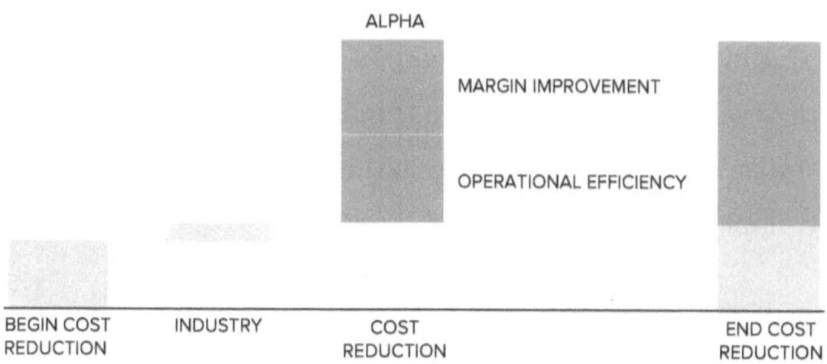

System Integration

When AI is deployed for margin improvement and operational efficiency, the AIs typically connect with the systems of record that hold operational, financial and transactional truth. This includes the primary business system and ERP (Enterprise Resource Planning) as well as the HRM (Human Resource Management) system, SCM (Supply Chain Management), MES (Manufacturing Execution System) and scheduling modules.

MARGIN IMPROVEMENT

Margin improvement involves enhancing profitability by spending less on costs. It is a key cost reduction lever because it directly increases profitability, enabling companies to reinvest in growth initiatives or to pay down debt.

Traditional Approaches

Traditional margin improvement relies on manual budget reviews, periodic cost audits, renegotiating supplier contracts, workforce reductions and broad-based expense control measures.

How AI Can Help

Key cost examples for AI margin improvement typically include areas such as supply chain optimization, predictive maintenance, workforce scheduling, inventory optimization and energy management. Whatever the area of cost reduction, AI margin improvement helps by enabling precise identification and ongoing optimization of fixed and variable costs.

Fixed Cost Reduction

AI enhances productivity by optimizing resource utilization and streamlining processes. AI can identify redundant and less effective overhead spending and then replace those parts of the business at a lower cost without impacting business operations.

Variable Cost Reduction

AI optimizes efficiency and streamlines the cost of goods and services by identifying and improving production processes and material usage.

Client Case Study: AI Sales Agents

A growth equity–backed online marketplace for recreational boats and vehicles was spending a disproportionate amount on customer acquisition—getting listings on its marketplace. The board tasked the CEO with looking into AI to see if it could be used to help improve margins.

To reduce fixed costs, high-volume sales jobs that were highly repetitive, were rules-based and required looking through large volumes of data were identified. The first AI sales development agent scored leads based on their likelihood to convert from being marketing qualified leads (MQLs) to sales qualified leads (SQLs).

A free trial of the AI sales agent demonstrated it was more effective in choosing the leads to pursue than its human counterparts and cost less to hire. This prompted our client to steadily "hire" and roll out new AI agent types to augment or replace more sales and marketing employees.

Value Creation Framework

MARGIN IMPROVEMENT AI AGENT

VALUE
Company: Get more listings with less headcount.
↓
USEFUL INFO
A list of leads scored based on their probability of converting.
↓
AI MODEL
CREATE
CREATE list by triangulating all data.
↓
DATA
Past MQL to SQL conversion data (RAG),
ICP data (business system, internet, social media).

OPERATIONAL EFFICIENCY

Operational efficiency involves enhancing profitability by optimizing business processes, resources and workflows. Whereas margin improvement is about spending less, operational efficiency is about doing things better. Like margin improvement, it is a key cost reduction lever because it also directly increases profitability, enabling a reinvestment in growth initiatives or payment of debt.

Traditional Approaches

Traditional operational efficiency relies on manual process audits, lean management techniques, Six Sigma methodologies and workflow analysis.

How AI Can Help

AI operational efficiency helps by automating and optimizing processes in both asset-heavy and asset-light businesses.

AI can play a major role in evolving maintenance. Most businesses today employ reactive maintenance, preventative maintenance, or proactive maintenance (Figure 15).

Figure 15 **Maintenance Evolution from Using AI**

REACTIVE → PREVENTATIVE → REACTIVE
PREDICTIVE → PRESCRIPTIVE

A fan favorite in asset-heavy businesses is AI-driven predictive maintenance. It identifies equipment failures and downtimes before they occur, enabling preventative actions to minimize disruptions. It does this by matching pre-learned failure signatures to current data. When there's a match, it makes a prediction that there will be another failure like the one that happened in the past.

Prescriptive maintenance uses AI to go one step further. After finding the failure match, it'll reverse engineer the problem and make adjustments so that the problem doesn't ever occur in the future.

Optimizing processes is another strength of AI. Once the process is codified into an AI model, the model learns how to be more efficient over time.

Although operational efficiency has traditionally been associated with asset-heavy businesses, with AI it can also be applied to service-based and tech companies.

In software companies, for example, AI can be used to accelerate development by writing code, automating coding tasks and suggesting improvements. It can also reduce overall development cycles by identifying bugs faster and more accurately than manual testing.

Client Case Study: Improving Factory Performance

A sponsor-owned precision manufacturer of special-purpose gaskets and seals for the aerospace industry had a difficult time meeting demand. To keep up, the company was buying more equipment and outsourcing business.

Internal capacity was increased after integrating an AI-based performance monitoring system to sense, display and suggest improvements about the manufacturing process. We did this by tuning the AI models of an off-the-shelf Overall Equipment Effectiveness (OEE) manufacturing product to represent the gasket and seals manufacturing process.

Using AI to optimize the manufacturing process saved the client from buying new machines and reduced outsourcing costs, resulting in lower CAPEX and higher margins.

Value Creation Framework

OPERATIONAL EFFICIENCY AI
VALUE Company: Lower fixed manufacturing costs. ↓ USEFUL INFO OEE parameter tuning for key machines and a schedule that optimizes asset utilization. ↓ AI MODEL DETECT ANOMALIES → PREDICT → OPTIMIZE DETECT ANOMALIES in machine operations; then PREDICT when a machine is going to fail; then OPTIMIZE schedule based on potential failures and line load changes. ↓ DATA Visual data (cameras), temperature (IoT), vibration (IoT), electric current (IoT).

CHAPTER 9

MULTIPLE EXPANSION OPPORTUNITIES

Multiple expansion refers to increasing a company's valuation multiple by means beyond operational improvements. A higher valuation comp (comparable company multiple) directly influences valuation.

In this chapter we'll discuss how AI can help a business expand through inorganic growth (add-on acquisitions) and make structural changes so it's comparable to more valuable types of companies (multiple rerate) (Figure 16).

Figure 16 **Multiple Expansion Value Levers**

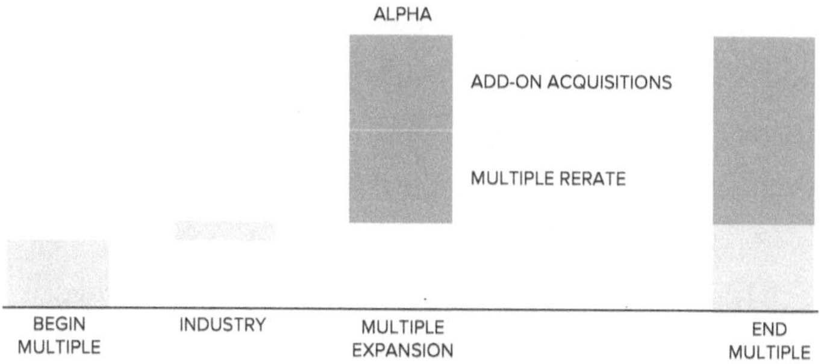

ADD-ON ACQUISITIONS

Add-on acquisitions involve expanding a company's valuation multiple by strategically acquiring smaller businesses that complement or enhance the platform company's operations and opportunities. It is a key multiple expansion lever because it accelerates growth and expands market reach beyond organic initiatives alone.

Traditional Approaches

Traditional add-on acquisitions rely on commercial M&A databases, manual market scans, plumbing personal networks and template-based diligence processes and involve investment bankers or corporate development teams.

How AI Can Help

AI add-on acquisitions can support the entire M&A process, but here we'll focus on data, or more specifically, missing data. From a data perspective, this AI value lever is the last to consider—pre- or post-acquisition—because it can boost and help make other already-identified AI value levers more impactful. It does this by zeroing in on acquiring the missing proprietary data needed by the other levers. The steps are:

1. Identify the (AI) value levers to be implemented as part of the platform company's value creation plan.
2. Consider the proprietary data needed for each lever based on the chosen KPI.

3. Do a gap analysis between the data the platform company has and needs to implement its value creation plan.
4. Use AI-driven analytics to evaluate add-on acquisition targets based on these data gaps.
5. Purchase from this list of rank-stacked businesses, underwriting the offer prices by considering not only the roll-up effect on the platform's value but also the boost the add-on businesses' data will give to the platform company's (AI) value creation plan.

Client Case Study: Data Acquisition

The same sponsor-owned OEM of midstream gas compression equipment from the customer retention case study wanted to roll up smaller competing service providers in the Permian Basin. Besides wanting to reduce part replacement churn by acquiring maintenance and repair shops closer to its customers, it also wanted to expand its high-margin repair business by supplying parts to the owners of competing gas compressors.

Acquisition targets were identified by geographic location and by the data the company had on competitor compressors and their machine locations. We did this by customizing an off-the-shelf AI analytics system that scoured public data, our proprietary data and purchased databases to identify the gaps in needed market expansion data.

This helped our client identify the maintenance and repair shops that were in the right geography and also had the market expansion data it was looking for.

Value Creation Framework

ADD-ON ACQUISITION AI
VALUE Company: A more impactful roll-up strategy. ↓ USEFUL INFO A list of roll-up targets in a geographic location that will help reduce churn and have proprietary data that will enable market expansion. ↓ AI MODEL CREATE CREATE an analysis of the individual acquisition targets and their proprietary data. ↓ DATA Market expansion data (internet, purchased databases), like acquisition targets' ICP (business system, internet), acquisition targets' locations (maps).

MULTIPLE RERATE

Multiple rerates involve expanding a company's valuation multiple with strategic repositioning supported by product and structural changes that will improve the future perception of the company. It is a key multiple expansion lever because it boosts enterprise value independent of EBITDA growth.

Traditional Approaches

Traditional multiple rerates involve restructuring initiatives, brand changes and industry benchmarking.

How AI Can Help

AI multiple rerates help by changing the comp of the company from its current peers to companies that are more technically advanced and yield higher multiples.

Public and private markets value businesses that incorporate AI into their products (services, places) and operations more than businesses that don't. AI companies are valued higher than software and other types of digital companies, which are valued higher than tech-enabled companies, which are traditionally valued higher than traditional physical companies (Figure 17).

Figure 17 **Relative Valuations by Company Type**

TRADITIONAL < TECH-ENABLED < SOFTWARE < AI

⟶

VALUE MIGRATION

We want some of this "AI juice" for all our portfolio companies because unlocking the AI premium will attract a higher valuation at sale from both strategic and secondary buyers. For example, publicly traded companies with AI products or strategies trade at about twice the forward multiple, or valuation multiple, of non-AI peers (Tomasz Tunguz,

2024). Higher multiples for AI businesses reflect expectations of faster growth, stronger moats and scalability.

This premium comes from making AI structural changes and, when possible, using AI to support higher-valued business models.

AI Structural Changes

The structural changes that support an AI rerate are a by-product of implementing any of the other AI value levers. Typically, a company's internal AI human resources are not enough to execute most of the AI value levers covered here. External AI expert resources then are employed to develop, document and then transfer the IP to the company. During this process an in-house AI team must be trained or hired. Structural changes to the company come from having AI IP intangible assets and the internal AI resources and know-how required to further grow that IP.

AI Business Models

New business models too are a by-product of implementing another AI value lever—specifically AI Price Optimization. As explained in the pricing optimization section of Chapter 7, recurring business models are valued higher than one-off sales. For example, subscription- and consumption-based models tend to create more predictable revenue streams, which can command higher multiples. Furthermore, being first to market with a service or outcome business model can be disruptive.

Structural changes and new business models are highly desirable because both produce a valuation double dip by improving both EBITDA and the valuation multiple.

Client Case Study: Expanding Company IP

A sponsor-owned manufacturer of extrusion equipment had received customer complaints about being responsible for production downtime and material wastage due to their forming punch not being properly aimed. This error occurred because the process was done by eye and the equipment was adjusted by wrench.

We added sensors to the extrusion body and digitized the adjustment mechanism so AI could better target the forming punch. We did this by tuning an AI analytics program's model to take aim based on equalizing the force data along the extrusion path.

Although the value proposition discussed in this last case study was for the end customer, value was also produced for the company. Expanding the company's IP from mechanical engineering to software using AI resulted in structural changes that should migrate the company's comp at sale from that of an engineering company toward that of an AI company with a higher valuation multiple.

Value Creation Framework

MULTIPLE RERATE AI

VALUE
End customer: Less material waste and less machine downtime.

↓

USEFUL INFO
Ideal position and angle to target the forming punch.

↓

AI MODEL
OPTIMIZE
OPTIMIZE the underconstrained equation based on extrusion parameters.

↓

DATA
Extrusion force data (IoT), targeting position and angle mechanism (IoT).

• • •

In Part Two we've shown that your AI strategy consists of one or more AIs whose value can be classified and quantified based on the type of value lever they enable. For example, your AI strategy could consist of three AIs, one each to expand market share, reduce churn and improve operational efficiency.

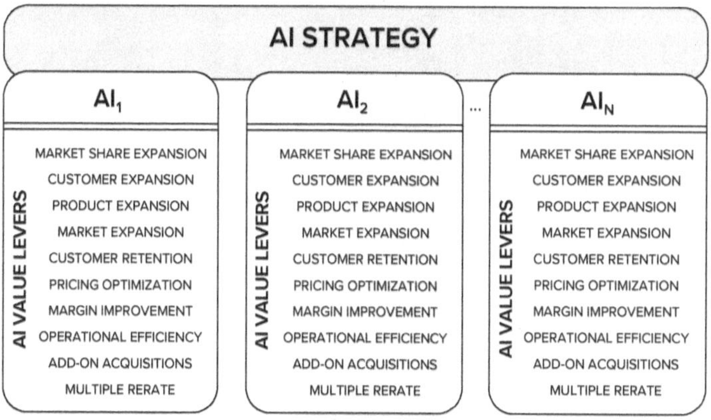

Considering value creation from a value lever perspective is an alternative to an AI Impact Assessment to identify the AIs in your strategy.

In Part Three we'll go through the AI Tech Stack so you can talk confidently with all AI stakeholders. This is especially important when working with your tech team to ensure that they don't stray from delivering the defined value during development and that they deliver on time and on budget.

PART THREE

EXECUTING YOUR AI STRATEGY

As discussed in the two previous parts of this book, a company's AI strategy manifests as one or more value-creating AIs. Earlier we learned what AI was able to do and how to use it to create value. This last part of the book explains AI tech—the nuts and bolts of how it works including a detailed example of the math and tech involved.

Your value-creating AIs will be in the form of off-the-shelf libraries or applications. The work is not in developing models but in choosing which models to use, training these models with proprietary data and then integrating them into the company's products (services, places) or operations.

To do this, resources will be needed. If your company is not staffed with software and data science talent, some form of external team will be needed to kick things off. Whether internal or external, these teams will be filled with techies, so they'll do what techies naturally do: build technically beautiful AIs. But technical beauty is almost always different from value beauty. If tech teams are left to their own devices, even with the best of intentions, things can go awry quickly. To keep things in line, this part of the book will also give you the background you need to ensure that the tech team builds a thing of value beauty aligned to your AI strategy.

Coming up you'll up your value creation game by learning the AI Tech Stack to understand what AI is and how its applications and agents are built. Not enough to program it yourself, mind you, but well enough to discuss your AI investment with all stakeholders involved.

If this part seems technical, it's because it is. It's why we need data scientists. The goal here is to give you a sense of what needs to be developed, so skimming, instead of reading, through the super-techie stuff is okay because both roads lead to the all-important AI Requirements Doc.

CHAPTER 10

THE AI TECH STACK

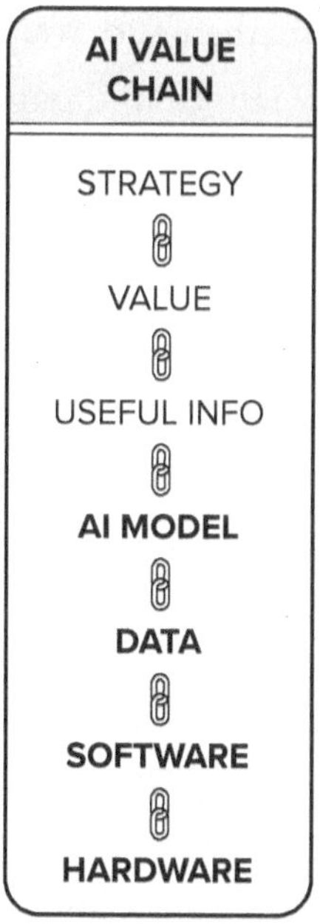

On the bottom half of the AI Value Chain hang the four layers of the AI Tech Stack: the AI Model, Data, Software and Hardware.

Previously we discussed the AI model and its data in the context of the AI Value Creation Framework. This time we'll look at them from a tech and math perspective. Let's go through each layer of the tech stack, one at a time, starting with the AI model. Then, near the end of the chapter, the road roller case study will return as a real-world example of how the tech and math were used to create value.

AI MODEL

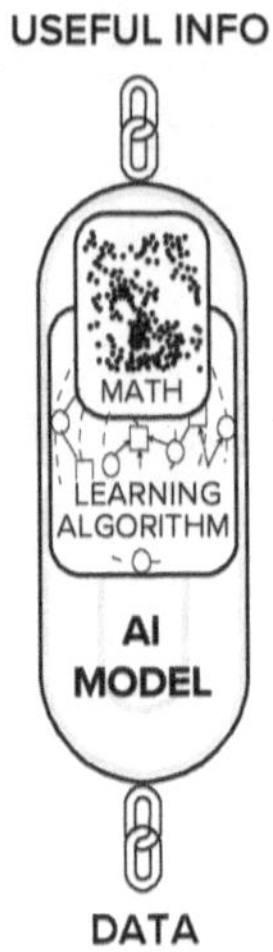

When we talk AI, what we're really discussing, at least in the context of value creation, is the AI model. Why?

All AI value comes from transforming data into useful information.

The AI model is the transformer—it mathematically transforms the ingested data into outputted useful information. But how?

Technically, the anatomy of an AI model is mathematics wrapped in a learning algorithm wrapped in Input/Output (I/O) software.

It is the second part, the learning algorithm, that differentiates AI models from all other types of models, but first let's look at the math.

Mathematics

At a high level, an AI model's math can be described as:

$$\text{USEFUL INFO} = f(\text{DATA})$$

Where f is the mathematical function used by the AI to transform data into useful information. The learning algorithm then adjusts f until the transformation produces the desired output.

The type of mathematical functions used by AI models to transform data are primarily probability and statistics, linear algebra, calculus and discrete mathematics (Figure 18).

Figure 18 **Types of Math Used in AI Models**

Probability and Statistics: Bayesian, Markov chains, fitting

Linear Algebra: Matrix math, vector spaces, tensor operations

Calculus: Differential, integral and variational calculus

Discrete Mathematics: Graph theory, Boolean logic

Learning Algorithm

Before any AI model can be used, it needs to be trained to produce the output we want (Figure 19). During training, the learning algorithm tunes the model by adjusting the parameters in its mathematical equation dynamically to minimize the fitting error to the training data. Error in this context means the difference between the model's predictions and the expected results.

Figure 19 **AI Model Training Steps**

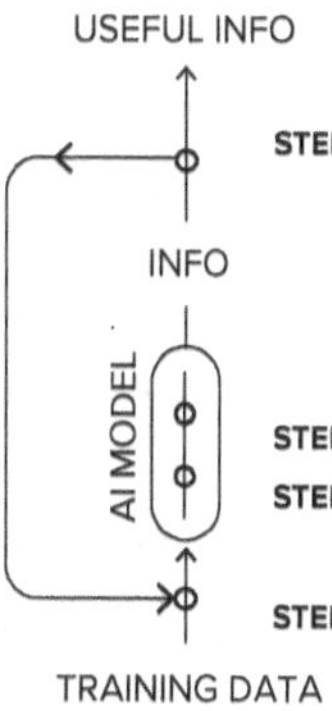

STEP 4: Outputted information is checked for accuracy. If the output is deemed accurate enough, the training is complete, and the outputted information is useful. Otherwise the training data is once again inputted back into the model.

STEP 3: Math equation transforms data into information.

STEP 2: Learning algorithm adjusts equation's parameters.

STEP 1: Training data is inputted into the model.

Training is an iterative loop of adjusting parameters to minimize error. Once the error has been minimized to an acceptable threshold, training is complete, and the outputted info is useful.

The four main types of learning algorithms used to train AI models are supervised learning, unsupervised learning, reinforcement learning and deep learning. Each algorithm uses a different technique to adjust model parameters and quantify error (Figure 20).

Figure 20 **Types of Learning Algorithms Used in AI Models**

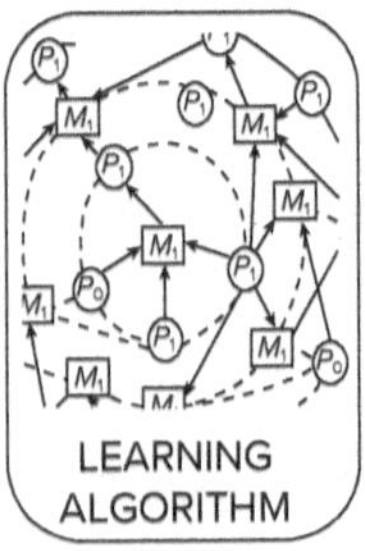

Supervised Learning: Regression, SVMs, neural networks

Unsupervised Learning: K-means, principal component analysis

Reinforcement Learning: Q-learning, neural networks

Deep Learning: GenAI, neural networks

The AI models themselves are widely available, but their proprietary value lies in how they are trained. (For a full list of commercial AI model types, see Appendix 1 at the end of Chapter 1.) No one in our space invents new models; off-the-shelf software is always used. Therefore, a company's AI IP isn't the AI model itself. The types of IP to cultivate are the models selected, the model's parameter values and their workflow, the know-how used for training models, how AI is integrated into company products (services, places) and operations, and the company's proprietary data.

Inferencing

Once the learning phase is complete, that is, once the parameters of the model's equation are set, it's ready to be used in production and no longer uses the learning algorithm.

In production, AI models are like any other—data is inputted into the model and transformed by its mathematical equations to output the answer.

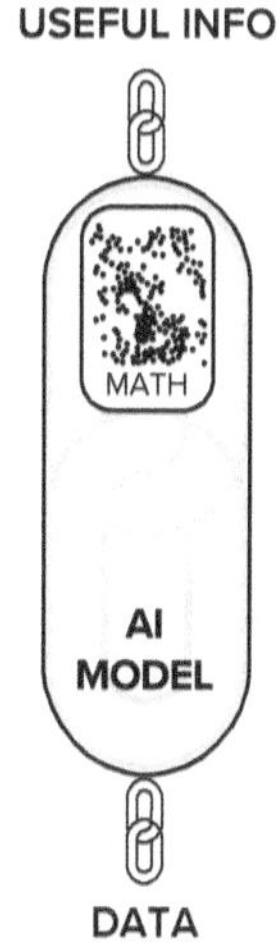

Input → Transformation → Output

After training, the process of transforming the input into the output is called inferencing. The outputted useful information is then passed on to the company's product or operations.

And because inference is much faster than training, it's generally computed locally (on the edge of the network) rather than in the cloud. This is how all AI works: inputted data is mathematically transformed by the AI model into information.

For example, consider GenAI chatbots. You input a prompt that is then transformed by the CREATE AI ability, and the answer is outputted:

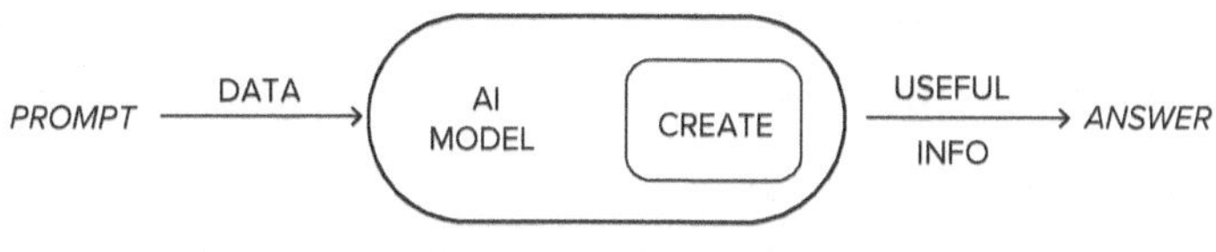

In the road roller example, the application inputs HMA temp, drum weight, drum frequency and roller speed; that data is transformed by the SIMULATE and OPTIMIZE AI abilities; and the speed for the road roller is outputted:

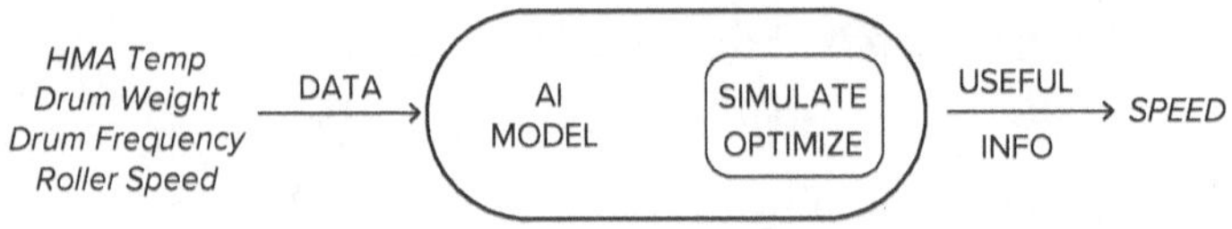

DATA

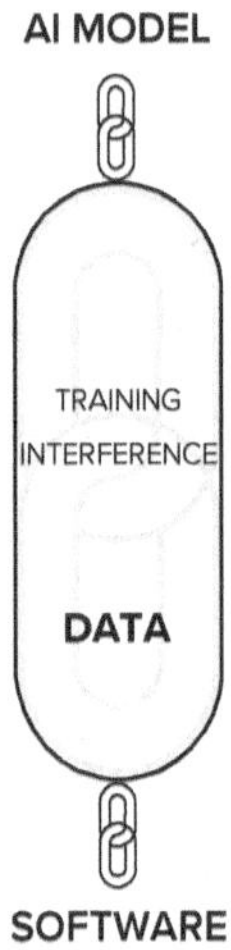

People no longer say "big data," because these days all data is big data. And the biggest data of all is the training data for the foundational GenAI models, which is basically . . . all the data of the entire internet! They may not say "big data," but people do say, "Data is the new oil," implying that companies are sitting on huge untapped resources that are just waiting to be monetized. In the context of meaningful value creation, however, this just isn't true. Successful companies determine the data they need, and if they don't have it, they make or find it.

Surveying the available data in your company can give you some ideas, but as we've seen with the AI Value Chain, data driving strategy is backward thinking. Instead, companies must produce the data their AI models need to create value.

Data IP is the proprietary data the company collects from its customers, products, employees and operations. If this data is not proprietary, it's not intellectual property—nor will it produce a unique competitive advantage.

Let's look at the two main types of data: training data and inference data.

Training Data

Training data, as the name suggests, is the input used to teach AI models. Preparing it—extracting, standardizing scales and units, and loading it into the model—can consume up to 75% of a data scientist's time. This overhead must be anticipated and planned for. Fortunately, the landscape is shifting. In an example of AI assisting AI, some practitioners are now using Generative AI to automate this grunt work: transforming messy, unstructured data into clean, structured formats ready for model ingestion. While tedious, that's not the hard part. The hard part is how a company trains its models. This training "pipeline" is the secret sauce. If an outside service provider is hired for initial models training, it is very important that this intellectual property is transferred to the company.

Inference Data

Inference data, or just data, is the data fed into the AI model after it's been trained. The data is accessed from company datastores, IoT sensors and business systems. The main inference data stream is (or should be) proprietary, but publicly available data, such as, say, weather conditions, is often used as a data stream flowing into the AI model.

SOFTWARE

Software's job, when it comes to AI, is to collect the AI's data, prepare it for inference and then route it into the AI model. The software that holds the AI model is packaged as AI applications and AI agents (see Chapter 5). Most will be off-the-shelf, including the plumbing software that integrates the AIs into the company's products or operations, but some "glue" may have to be custom written by a developer or an AI.

Another category of software is the digital twin. As discussed in Chapter 5, the digital twin consists of a software wrapper around an AI model. Together they digitally represent how the physical product creates value. By being software-defined and data-driven, the physical product, through its digital twin, can use AI applications and agents the same way software products do.

All software, whether apps/agents, digital twins or integration code, exists so the AI models can create value.

HARDWARE

Hardware is deployed both locally, on the edge in networking terms, and remotely in the cloud. Edge hardware is typically used for inference, while cloud infrastructure is generally reserved for training workloads.

Local hardware may need to be purchased outright, but cloud infrastructure is best rented rather than bought due to its scalability, cost-efficiency and support requirements.

Whether operating on the edge or in the cloud, hardware serves three core functions: compute, storage and networking.

- Computer hardware powers both the training of AI models and their execution during inference.
- Storage hardware holds datasets—both raw and processed. Data consumes the most space by far, but additional capacity is allocated for trained models and logs, which support reproducibility and monitoring.
- Networking hardware transfers data from its source to storage systems, and from storage to the compute environment where models reside. Bandwidth demands are highest during data ingestion and inference.

When using AI for physical products, proprietary product or operational data is captured by sensors and transmitted by IoT.

The fidelity of the captured data must be chosen because the more data collected, the higher the costs of IoT, networking, storage and compute hardware.

Overall, the purpose of the tech stack is to move data from the hardware to the AI model so it can be transformed into useful information.

While many are familiar with LLMs and generative AI, this chapter and this book deliberately highlight traditional Analytical AI and non-generative neural networks. The message is clear: Value creation in AI isn't limited to the flashy billion-dollar models we continual hear about. Simpler, well-targeted approaches often outperform GenAI in real-world deployments, delivering a faster ROI, lower operational risk and greater scalability.

TECH STACK EXAMPLE FROM CASE STUDY

Let's geek out a little by going back to the road roller case study first discussed in Chapter 2 and go through a simplified example of the tech. If you remember, we developed an AI model for my client's OEM road roller customer that simulated asphalt compression. The common links on the AI Value Chain for the road roller case study and this example are the AI model and its data. Whereas the value creation framework case study focused on how to define the AI model and data from a value creation perspective, here the focus is on how the AI application to control the road roller is made—the technology underpinning the AI model, including its data, software and hardware, in order to understand how they work within the AI Tech Stack.

AI Model

Math

As was shown in Chapter 2, we can rearrange the asphalt density function to solve for roller speed, resulting in the equation:

Speed = *f*(desired asphalt density, drum weight, drum frequency, HMA temp)

For this example, let's say that since the desired asphalt density, drum weight and drum frequency are known and unchanging, we can collapse them into one constant, *b*.

Our simplified function can now be expressed as:

Speed = *f*(HMA temp, *b*)

And for teaching purposes, let's further assume that the relationship between speed (*S*) and the HMA temp (*X*) is linear, resulting in the linear equation:

$$S \text{ [or Speed]} = m * X + b$$

where m = slope parameter

X = HMA temperature

b = constant combining desired asphalt density drum weight and drum vibration frequency

Look familiar? It's the equation of a line. Something you've undoubtedly used to graph data in Excel. The math used here is the same used by Excel.

For illustrative purposes, we're reducing the fidelity of the speed model to two dimensions and then doing linear regression. In practice, though, this is a multidimensional problem solved by non-linear regression (another mathematical method could be used, but the concepts explained here are the same).

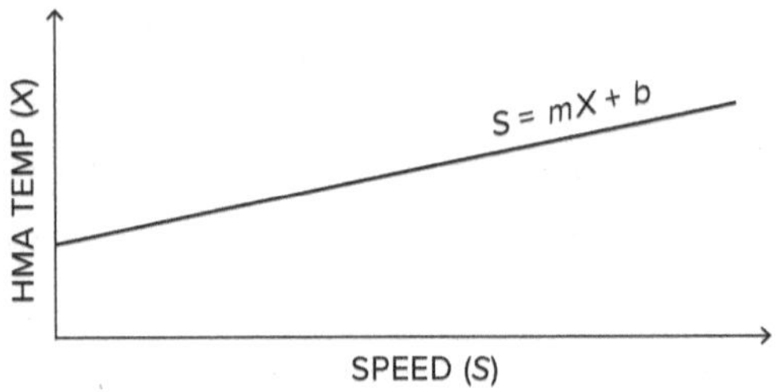

We now relate the speed of the road roller (*S*) to the temperature of the hot mixed asphalt (*X*) with a line. We can already see how the model is going to work, and it seems intuitive. At low HMA temperatures the roller needs to move slowly to effectively squish the cooler asphalt to the desired density, but at high temperatures the roller can move faster to compress the hotter asphalt because it's more malleable.

Learning Algorithm

Now to the AI. Remember, AI is math wrapped in a learning algorithm. What changes linear regression, the statistical equation, into linear regression, the AI model, is that the AI model learns how to best fit the data using a learning algorithm. To tie this to Chapter 2, it is the OPTIMIZE atomic AI ability that does the learning to best fit the equation to the training data.

To improve our understanding of the relationship between the roller speed (dependent variable) and HMA temp (independent variable), the learning algorithm tweaks the parameters of the equation to make it more accurate. Or said another way, the better this AI model fits the training data, the more accurate it is in predicting roller speed. In our linear regression AI model, we have two parameters that can be tweaked to improve fit.

The algorithm can change *m*, which changes the line's angle:

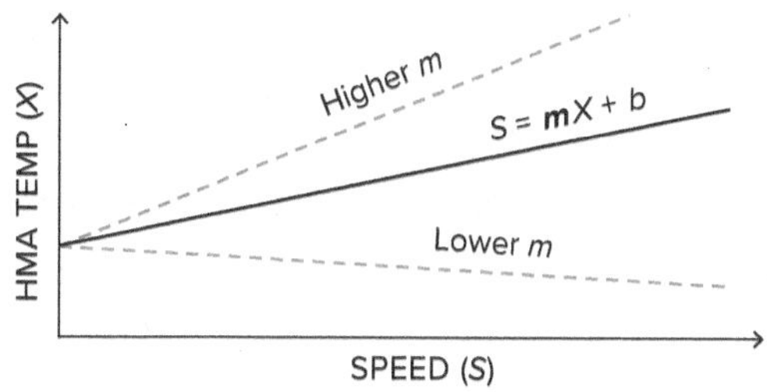

And it can change *b*, which changes the height of the line:

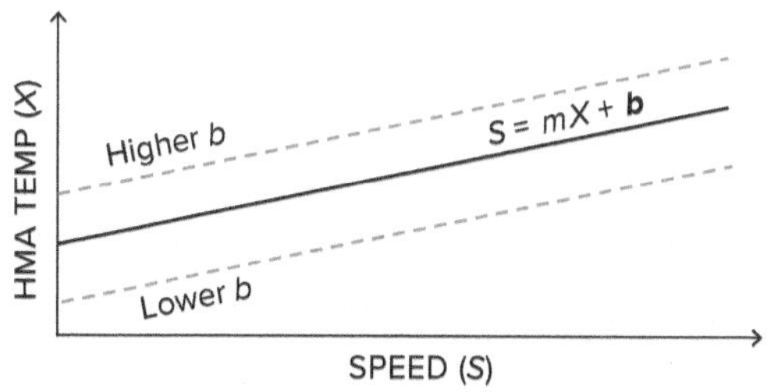

Note that to raise or lower *b*, we multiply it by the parameter *n*, which still results in a constant that we'll call *nb*.

Our linear regression AI model, $S = mX + nb$, learns from the data by trying different values of m and n to improve fit, which is measured as a score.

This is how all AI models are "trained." From simple linear regression to highly complex GenAI neural networks, the process is the same: Keep adjusting the model's parameters until the model best fits the data. When you hear, for example, that OpenAI spent months and hundreds of millions or billions of dollars training their latest model, this is exactly what's happening. It costs that much in time and money because instead of tweaking two parameter as is demonstrated here, their training involves tweaking billions of parameters, often called weights.

As mentioned before, although GenAI can do pretty much everything Analytical AI can do, there are many cases where it is just overkill. This is one such example. Using machine learning is more efficient than using GenAI by solving the problem analytically.

Data

Training

For our AI model, $S = mX + nb$, to learn how to be accurate, it needs to be trained by using training data.

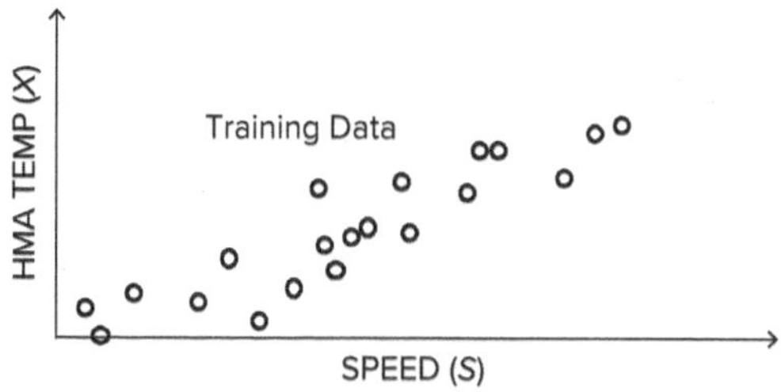

That is, the learning model changes the parameters *m* and *n* until the equation fits the training dataset best. The AI model tries all the combinations of *m* and *n* until it finds the highest fit score.

For example, based on this training data, equation 1 scored 51% and equation 2 scored 85%.

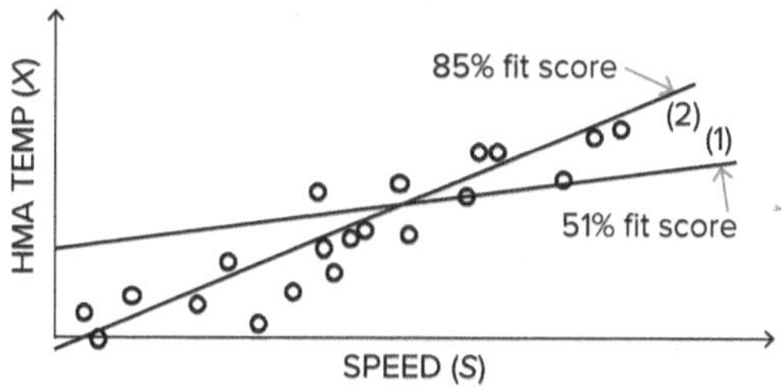

The only difference between the two equations is the values of the equation's parameters *m* and *n*.

Since we don't have the training data, let's use actual road roller data and work backward to get realistic parameters for our linear regression equation. Based on age-old engineering tables, we know HMA temperatures for a specific environment range from 225°F to 350°F and road roller speed ranges from 1 mph to 8 mph.

Using linear algebra on this training data, we get parameters $m = 7/125$ and $nb = -11.6$, resulting in the equation:

$$S = 7/125X - 11.6$$

which we can see in this plot:

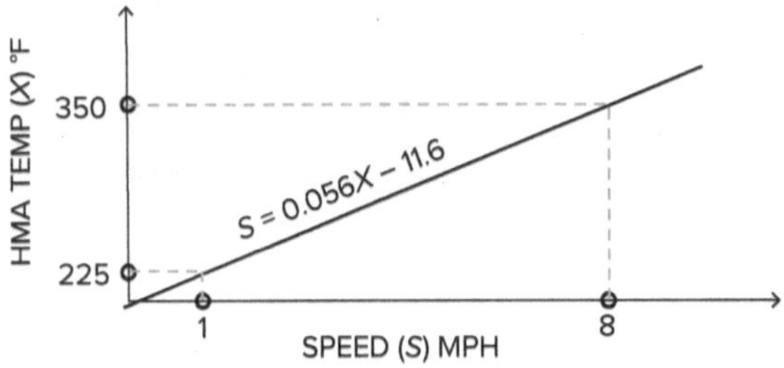

Inference

Now that the parameters (weights) *m* and *b* of our road roller's speed model have been determined through training, we can input a new dataset (*X*) into it to output the modeled speed of the road roller (*S*).

In the case of the road roller, the data we use to inference is the temperature of the HMA (*X*) that is delivered from the temperature sensor(s) on the road roller.

For example, if the HMA temperature is $X = 300$°F, then the max speed the road roller can travel is $S = 5.2$ mph. Over time the HMA will cool. So then if the temperature is $X = 250$°F, the speed of the roller will need to slow down to $S = 2.4$ mph to properly compress the cooler HMA. This is AI inferencing.

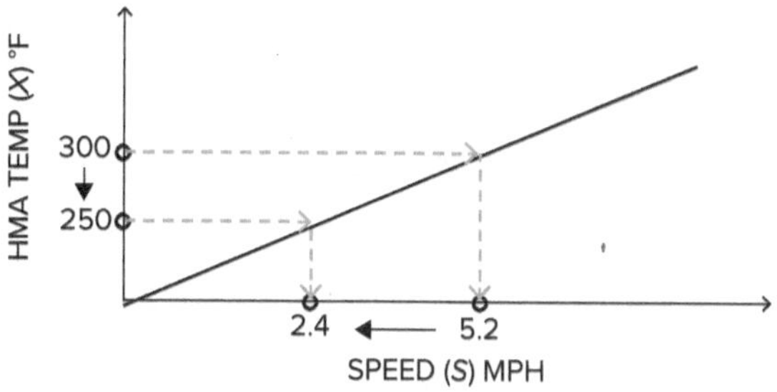

Software

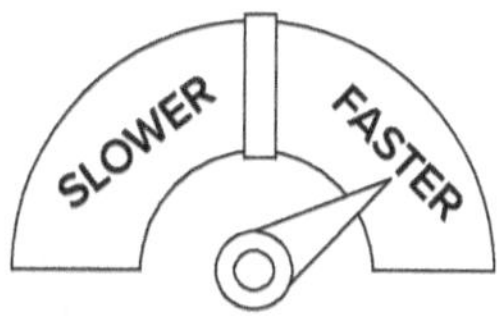

In this example, the AI app for the road roller reads the temperature of the cooling HMA from its sensors, inferences it by inputting *X* (temperature) into the AI model and then compares the outputted *Y* (speed) with the current actual speed of the roller. If the output of the AI model is higher, it adjusts the speed meter to instruct the road roller operator to slow down. If the output is lower than the current speed, the software adjusts the speed meter to instruct the road roller operator to speed up. To speed up or to slow down is the useful information produced that makes this road roller more valuable.

Hardware

In this example the road roller's hardware includes one or more HMA temperature sensors, and depending on the network architecture, it has both local (on the road roller) and managed services for networking, storage and compute.

And again, different software and hardware are used to train the model.

The next step taken to bridge the completed strategy work to the execution of the road roller's AI application, or any other AI, is to define the AI Requirements Doc.

The AI Requirements Doc

Our job isn't to dictate how to develop the tech stack of our AIs. Our job is to be crystal clear on their requirements. Enter the AI Requirements Doc.

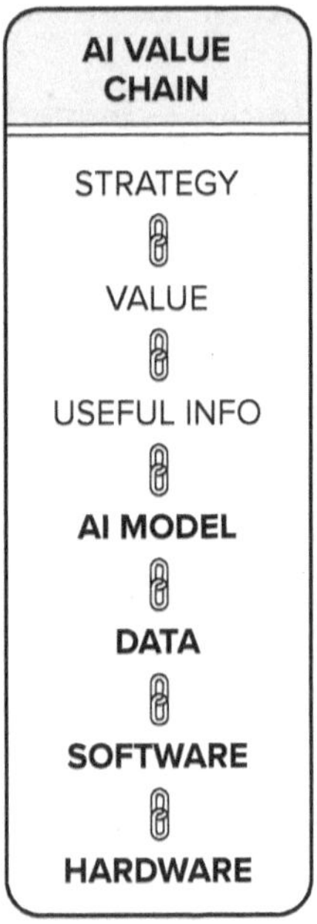

Since we are not developing the tech stack ourselves, the way we impose our will on the development process is by delivering a requirements document that prioritizes and preserves the value propositions of the AIs to be developed.

The AI Requirements Doc bridges the age-old chasm between the strategy team and the execution team. Without such a document, problems arise when using different entities to develop strategy and to realize that strategy. The AI Requirements Doc is our universal and necessary interface with any development partner.

To arrive here, at the precipice of the execution stage, means we used the AI Value Creation Framework to produce the requirements for our AI model and its data. To develop the requirements doc, we look at the entire tech stack, starting with the AI model again.

Each of the four layers defines the technical requirements of the layer below it, and each layer is dependent on the layer above it. The AI model is dependent on the required useful information, and the AI model defines the requirements for its data, and so on down to the hardware.

When working with a development partner, whether it's an internal team or an external service provider, vendor or subject matter expert, insist that the partner consider the development of each layer of the tech stack in the following order: AI Model, Data, Software and Hardware.

A common rookie mistake is to allow the managed services provider (addressing the hardware layer) to be chosen by your development partner before considering the AI model's technical requirements and the type of data it needs.

Another misconception is that AI is either magic or art and therefore doesn't need project management and MLOps/DevOps (machine learning and software development operations) best practices. AI is a software and engineering discipline like any other, and it demands the same level of rigor, structure, and accountability throughout the development process. Enforcing the AI Requirements Doc will help ensure that your development partner delivers the value you defined within the timeframe and price you agreed to.

Intellectual Property

It's worth repeating here that although the AI will be developed using off-the-shelf products and APIs, valuable intellectual property will be developed. This includes the AI model's parameters, the model's training process and the code used to integrate the AI into the company's products and operations. It also includes the proprietary data and metadata the company collects from its customers, products, employees and operations and how this data is organized and normalized in preparation for feeding it to the AI model.

Typically, outside help is needed to get started, so as part of the external development process, the IP must be transferred in-house, and an internal dev team, no matter how small, should be hired to maintain and further develop the AI.

This structural change of having an internal AI team improve existing AIs and building new ones contributes to what I call the AI rerate, which has been shown to double the valuation multiple (see Chapter 9) for when the time comes to sell the company.

Data Sovereignty and RAG

Related to IP—or more accurately, related to not leaking IP—is the concept of *data sovereignty*. Data sovereignty is about keeping your company's proprietary data under your control so it isn't exposed or used to train external GenAI models that competitors could access.

A Retrieval-Augmented Generation (RAG) system is a private company knowledge base that works with GenAI LLMs to combine general world knowledge with your proprietary data. This setup generates answers that are accurate, current and grounded in company-specific knowledge while maintaining data sovereignty.

• • •

As discussed earlier, an AI strategy consists of one or more AIs. Part Three has shown that from a value-preserving perspective, these apps and agents are best described through the AI Tech Stack and are developed by fulfilling the specifications outlined in the AI Requirements Document.

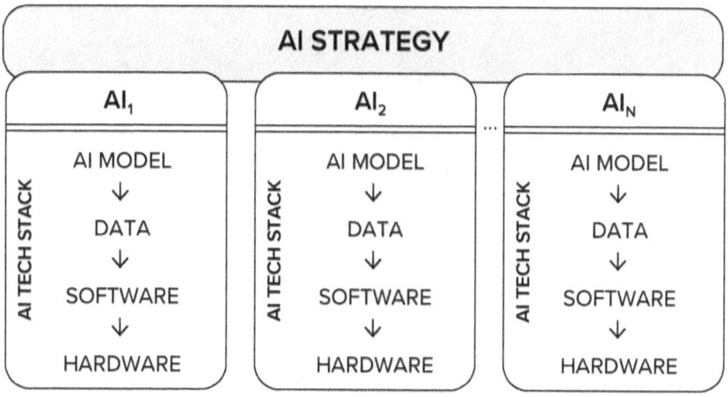

We'll conclude the book by showing how to get started with your first AI investment.

CONCLUSION

GETTING STARTED

It's possible that your firm has started using GenAI internally to get a "feel" for AI. Or maybe internal projects have been started to use a chatbot to quicken select firm operations. Awesome. But to get started at the portfolio level and really move the needle, thinking needs to shift to the six most important questions answered in this book.

1. HOW DO I DEVELOP MY AI STRATEGY?

An AI strategy is developed by selecting and defining one or more valuable AIs that advance the strategy of the company. Successful AIs will deliver a minimum 10x return on investment and a double-digit percent increase in enterprise value within the holding period.

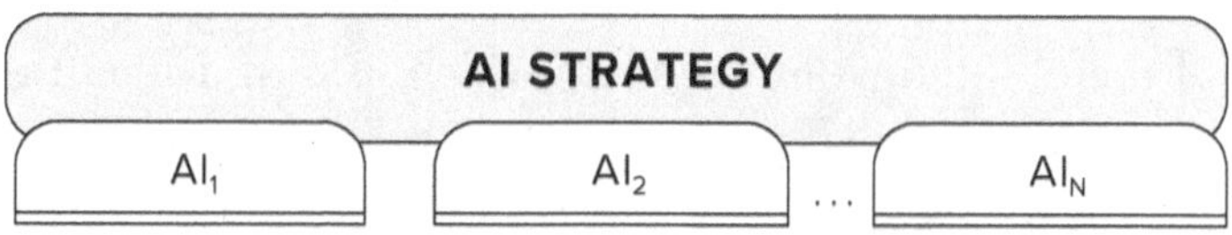

AI STRATEGY = one or more AI Applications or Agents

2. HOW DO I IDENTIFY AI RISK?

Like the opportunities that make it to the AI Value Creation Framework, risk is identified by doing an AI Impact Assessment (Figure 21). Impact is scored by the alignment of value propositions or opportunities at the task, operation, employee, business model, product and company level with AI abilities. Impacts to the right and up in the chart generally represent the bigger AI opportunities if implemented and the bigger risks if not implemented but a competitive company does so.

Figure 21 **AI Impact Assessment Template**

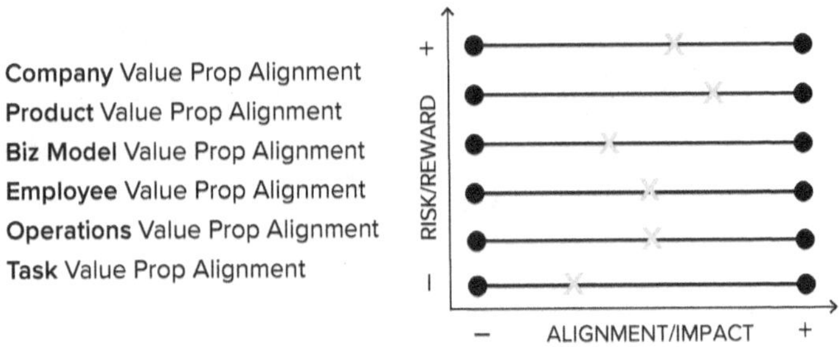

3. HOW DO I CREATE VALUE WITH AI?

Each AI is designed to create the desired value by using the AI Value Creation Framework. This value defines the type of useful information required, which in turn defines the output of the AI model that needs to be developed and the model's data that needs to be inputted.

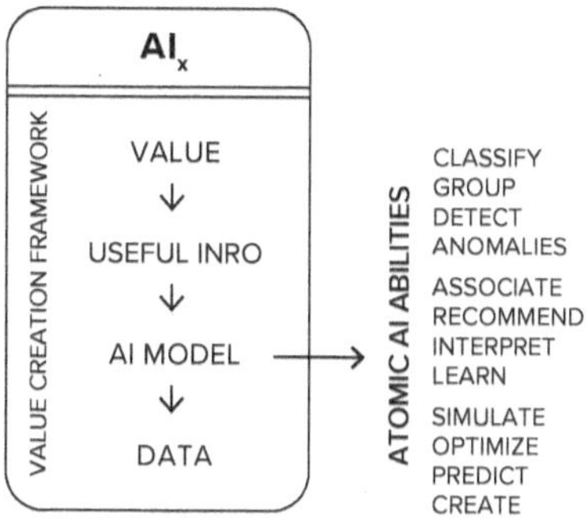

Each AI model's functionality is described by assembling one or more AI abilities together.

4. HOW DO I QUANTIFY AI VALUE?

Once defined, each AI's value is correlated to a value lever by the way it increases enterprise value. After being classified in this traditional way, it's straightforward to quantify each AI's value with conventional financial projections.

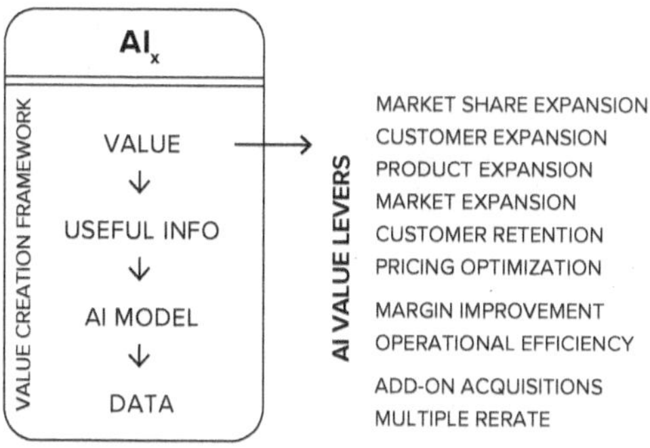

After their impact has been quantified, the AIs can be ranked with each other and other value creation options to determine which should ultimately be implemented.

5. HOW DO I EXECUTE MY AI STRATEGY?

Unless you're coding Python yourself, your AI strategy, consisting of one or more AIs, will be developed by an internal or external tech team. Either way, the AI Requirements Doc is used to bridge strategy with execution. The development of this document is guided by working through the AI Tech Stack requirements from top to bottom: AI MODEL → DATA → SOFTWARE → HARDWARE. Once complete, the AI Requirements Doc provides the information needed by your tech partners to accurately cost the AI investment.

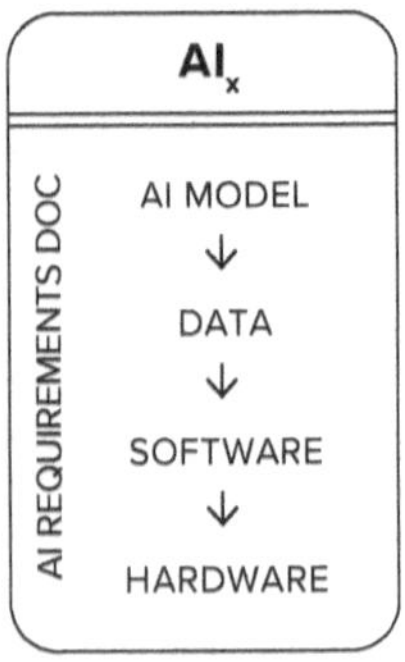

If you've made it this far, congratulations! You're prepared to get started. Perhaps while reading this book a few portfolio companies came to mind, or maybe you read a recent CIM (Confidential Information Memorandum) and saw an unmentioned AI value creation opportunity you thought would contribute to growth or cost reduction. That's great—it means you're ready. It means you're starting to see opportunities that were previously invisible.

AI is a broad horizontal layer that will permeate the products or services or operations of all companies. Software companies will be upgraded with AI, as will tech-enabled services companies; and through the digital twin, so too will all traditional companies that operate in the real world. How long will this take? It's unclear, but the time from now until AI becomes required table stakes represents your window of competitive advantage. Starting now, you'll be able to look at companies through the AI lens, pre- and post-acquisition, and with the help of this book you can incorporate AI into your investment theses.

Speaking of investment theses, that's what's next. The AI Investment Thesis is the answer to the last question:

6. HOW DO I GET STARTED?

The AI Investment Thesis generally kicks off the AI investment process. And if you think about it, this book has provided you with everything needed to develop it: identifying the investment opportunities, projecting their impact on enterprise value and costing their execution to get an ROI are the core components of any investment thesis and the starting point of the AI value creation plan.

The AI Investment Thesis is the first or second step of incorporating AI into your potential or existing investments. To put things into perspective, the overall AI investment cycle can be described by identify, plan and execute, along with its deliverables:

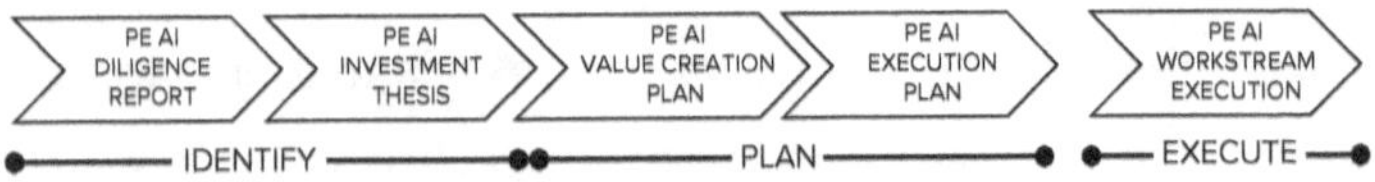

This book has primarily focused on the identify stage to get you started, but help is available if you want to keep going.

HELP IS AVAILABLE

If this book resonated with you and you decide your firm or company would benefit by going on this journey with a hands-on partner who's been there, done that, contact us at **aioperatingpartners.ai** for a free consultation. Or investigate our live and online training programs and our consulting and advisory services, which cover the entire AI investment cycle.

INDEX

Page numbers followed by *f* and *t* refer to figures and tables, respectively.

ABOUT THE AUTHOR

Bruce started his career as a mathematician and then programmer who quickly found his way to business through marketing. He worked his way through product marketing and brand management to be VP of Marketing for a large Microsoft subsidiary, and then he was the CEO of multiple tech companies, including an AI company. After that he put his operational experience to work as an operating partner for a middle-market private equity firm. He's currently the managing partner of AI Operating Partners, which advises and consults with PE firms and their portfolio companies.

Bruce is also the author of the McGraw-Hill—published book *IoT Inc.*, and the book *Digital Operating Partner*. His training courses on the digital transformation and the Internet of Things on LinkedIn Learning have certified over 300,000 professionals online.

Find out more at:

aioperatingpartners.ai

and

linkedin.com/in/bruce-a-sinclair/

www.ingramcontent.com/pod-product-compliance
Lightning Source LLC
LaVergne TN
LVHW090410210926
848480LV00002B/524